WIND EROSION IN NIGER

Developments in Plant and Soil Sciences

VOLUME 67

The titles published in this series are listed at the end of this volume.

WIND EROSION IN NIGER

Implications and Control Measures in a Millet-based Farming System

Edited by

B. BUERKERT, B.E. ALLISON, AND M. VON OPPEN

KLUWER ACADEMIC PUBLISHERS

DORDRECHT / BOSTON / LONDON

in cooperation with the

UNIVERSITY OF HOHENHEIM – STUTTGART – GERMANY

Library of Congress Cataloging-in-Publication Data

ISBN-13: 978-94-010-7222-9 e-ISBN-13: 978-94-009-1618-0
DOI: 10.1007/978-94-009-1618-0

Published by Kluwer Academic Publishers,
P.O. Box 17, 3300 AA Dordrecht, The Netherlands

Kluwer Academic Publishers incorporates
the publishing programmes of
D. Reidel, Martinus Nijhoff, Dr W. Junk and MTP Press.

Sold and distributed in the U.S.A. and Canada
by Kluwer Academic Publishers,
101 Philip Drive, Norwell, MA 02061, U.S.A.

In all other countries, sold and distributed
by Kluwer Academic Publishers Group,
P.O. Box 322, 3300 AH Dordrecht, The Netherlands

Cover photo: Sandstorm at the beginning of the rainy season.
Photograph by J.P.A. Lamers.

Printed on acid-free paper

Contents

PART 4 SOCIO-ECONOMIC EVALUATION OF WIND EROSION AND CONTROL MEASURES

PART 5 SUMMARY AND OUTLOOK

Preface

The West African Sahel is the transition zone between the Sahara desert in the north of Africa and the more humid Sudanian zones in the south. Although diverse in many ways, the Sahelian countries have the common problem of a fragile agricultural sector. This predicament is mainly caused by low inherent soil fertility, limited and unpredictable rainfall, frequent droughts, and wind erosion that accelerates soil degradation and desertification, compounded by rapidly growing populations. To assure food production in the future, means must be found to offset the trends of declining soil fertility and increasing soil degradation through wind erosion. This is a challenge for agricultural research.

Since 1985, the Special Research Program 308 'Adapted Farming in West Africa' at the University of Hohenheim in collaboration with the International Crops Research Institute for the Semi-Arid Tropics (ICRISAT) in Niger, has pursued the development of agricultural innovations for smallholder farmers in one of the most ecologically fragile regions of the world. The prevention of soil degradation, the restoration and maintenance of soil fertility, and the increase of land and labor productivity are key objectives of this multidisciplinary research program. From the beginning, a major focus of research has been wind erosion. Soil scientists, plant nutritionists, agronomists, economists, rural sociologists, and agro-climatologists worked together using different approaches and methods to better understand the implications of dust and wind-blown soil on soil properties and crops, as well as to evaluate erosion control measures such as windbreaks and mulching with millet crop residues in the millet-based farming system of Niger. Most research was conducted on-station at the ICRISAT Sahelian Center but with a farming systems perspective. From 1985 to 1987, a series of trials started with natural savannah windbreaks. In 1988, a tree windbreak trial was established. From 1990 to 1994, an intensive period of data collection on the effects of windbreaks and crop residues, as additional techniques for controlling wind erosion, took place. During this period, the idea for an interdisciplinary publication on wind erosion and its control was initiated and nursed by John Lamers, whose continuous interest greatly contributed to shaping the book during all stages of the publication process.

The book is divided into five parts. The first, an introduction to Niger (Chapters 1.1 to 1.3), addresses the climate, soil, and traditional farming systems as a prerequisite for evaluating research results presented in Parts 2 to 4. In Part 2, a review is presented on dust generation, accretion, and

B. Buerkert, B. E. Allison and M. von Oppen (eds.), Wind Erosion in Niger, ix–x.

the importance of deposited dust as a material influencing soil properties in the Sahel (Chapter 2.1). Chapter 2.2 focuses on data collected during strong connective storms at the beginning of the rainy season. The extent of wind erosion in terms of soil flux and damage to young millet plants is shown.

In Part 3, control measures for wind erosion such as soil ridging and natural savannah windbreaks (Chapter 3.1), planted tree windbreaks (Chapter 3.2), and a mulch of crop residues (Chapters 3.3 and 3.4) are evaluated and their effects on soil properties and dry matter production of protected crops are determined. Chapter 3.2 also considers the valuable products derived from the windbreak trees. Chapter 3.3 aims to distinguish between the nutritional effect and the erosion controlling effect of crop residues applied in high amounts; while Chapter 3.4 investigates different application levels of crop residues and their effectiveness in controlling wind erosion and improving millet biomass production.

Part 4 begins with a financial evaluation which integrates the agronomic, pastoral, and the silvicultural benefits of windbreaks (Chapter 4.1). Following this, the partial cost-benefit analysis of using crop residues as a mulch is presented. Linear programming techniques were used to estimate the comparative advantages of several mulching techniques for a typical farm in western Niger (Chapter 4.2). Finally, the evaluation of innovations and new technologies should not be restricted to agronomic and economic aspects but should also take the experience of the rural population into account. Therefore, the results of the crop residue and windbreak experiments are complemented with on-farm surveys that include the experience and knowledge of the rural population (Chapter 4.3). A summary of the conclusions drawn from the papers presented completes the publication and gives recommendations for future research and for the implementation of wind erosion control concepts by policy-makers.

This book is the first in a series of interdisciplinary publications of research conducted in West Africa by the University of Hohenheim in collaboration with national partners and international organizations. It is intended, in the first place, for researchers and institutions focusing on sustainable agriculture in the West African Sahel, as well as in others parts of the world where wind erosion affects agricultural production. Yet, targeted effort was made to ensure that this book be comprehensible for a broad readership including applied researchers, extension specialists, project leaders and students working in the field of wind erosion control, as well as national and international policymakers. Those involved in the planning and execution of natural resource protection programs, especially in West Africa, may also gain insights from this book.

B. Buerkert, B.E. Allison, and M. von Oppen
Stuttgart–Hohenheim, July 1995

Préface

Le Sahel constitue en Afrique de l'Ouest une zone de transition entre le désert du Sahara au nord, et la zone soudanienne plus humide au sud. Au-delà de leurs différences nombreuses, les pays du Sahel partagent le problème d'un secteur agricole fragile. Cette situation défavorable est en premier lieu le fruit d'une fertilité des sols intrinsèquement basse, de précipitations limitées et irrégulières, de sécheresses fréquentes et d'une érosion éolienne qui accélère la dégradation des sols et la désertification, ces facteurs se combinant avec une croissance démographique rapide. Pour assurer dans l'avenir une production alimentaire suffisante, on doit trouver les moyens de lutter contre le déclin tendanciel de la fertilité et la dégradation des sols due à l'érosion éolienne: ce défi relève de la recherche agronomique.

Le programme spécial de recherche 308 'Agriculture Adaptée en Afrique de l'Ouest' mené à l'Université de Hohenheim en collaboration avec l'Institut International de Recherche Agronomique pour les Zones Tropicales Semi-Arides (ICRISAT) au Niger, développe depuis 1985 des innovations destinées aux petits agriculteurs dans une des régions du monde les plus fragiles du point de vue écologique. La prévention de la dégradation des sols, la restauration et le maintien de la fertilité des sols et l'augmentation de la productivité de la terre et du travail sont les objectifs-clefs de ce programme de recherche multidisciplinaire. Depuis le début, la recherche s'est centrée avec force sur la question de l'érosion éolienne. Des spécialistes des sciences du sol, de la nutrition des plantes, des agronomes, des économistes, des sociologues ruraux et des agro-climatologues ont travaillé ensemble, mettant en œuvre des approches et des méthodes diversifiées visant à mieux comprendre l'influence de la poussière et des éléments de sol transporté par le vent sur les propriétés du sol et les cultures, ainsi que pour évaluer les méthodes de maîtrise de l'érosion telles que les brise-vent et le paillage de résidus de récoltes du mil dans le cadre des systèmes de cultures nigériens à base de mil. L'essentiel de la recherche a été conduit en station au Centre Sahélien de l'ICRISAT, mais dans une perspective de recherche-développement. Une première série d'essais de brise-vent naturels de savane ont débuté dans les années 1985–1987. En 1988, un essai de brise-vent arboré a été mis en place. Une phase de collecte intensive de données relatives à l'impact des brise-vent et des résidus de récoltes comme technique annexe de maîtrise de l'érosion éolienne, s'est déroulée de 1990 à 1994. C'est au cours de cette période qu'a germé l'idée d'une publication interdisciplinaire sur l'érosion éolienne à l'initiative de John Lamers qui en a pris en charge la supervision et dont l'intérêt soutenu

B. Buerkert, B. E. Allison and M. von Oppen (eds.), Wind Erosion in Niger, xi–xiii.

a grandement contribué à donner sa forme au livre à toutes les étapes du processus de publication.

Le livre s'organise en cinq parties dont la première, une introduction aux climats, sols et systèmes de production traditionnels du Niger (chapitres 1.1 à 1.3) constitue un pré-requis à l'évaluation des résultats de recherche présentés dans les parties 2 à 4. La seconde partie offre une synthèse sur les mécanismes de génèse et d'accumulation de la poussière et l'importance des dépôts de poussière comme matériau influant sur les propriétés des sols (chapitre 2.1). Le chapitre 2.2 est centré sur des données collectées pendant les forts orages de convection du début de la saison des pluies. L'ampleur de l'érosion éolienne en termes de flux de sol et de dommages sur les jeunes pousses de mil est mise en évidence.

Dans la troisième partie, les méthodes de maîtrise de l'érosion comme le billonnage et les brise-vent naturels de savane (chapitre 3.1), les brise-vent constitués par une plantation d'arbres (chapitre 3.2) et le paillage des résidus de récoltes (chapitres 3.3 et 3.4) sont évaluées et leur influence sur les propriétés du sol et la production de matière sèche des cultures protégées est déterminée. Le chapitre 3.2 prend aussi en compte les sous-produits valorisables des arbres brise-vent. Le chapitre 3.3 s'attache à distinguer l'impact des résidus de récoltes, appliqués en grande quantité, en termes de nutrition de la plante et de maîtrise de l'érosion, alors que le chapitre 3.4 évalue différents niveaux d'application des résidus de récoltes et leur efficacité sur la maîtrise de l'érosion éolienne et l'accroissement de la production de biomasse de mil.

La quatrième partie débute sur une évaluation financière qui intègre les bénéfices agronomiques, pastoraux et forestiers des brise-vent (chapitre 4.1). À la suite de cela une analyse partielle coût-bénéfice de l'utilisation comme mulch des résidus de récoltes est présentée. On a eu recours à des techniques de programmation linéaire pour estimer les avantages comparés de différentes techniques de paillage sur une exploitation-type de l'ouest du Niger (chapitre 4.2). En définitive, l'évaluation des innovations et des nouvelles technologies ne devrai pas se limiter aux aspects agronomiques et économiques et devrai prendre en compte l'expérience des populations rurales. Les résultats de l'utilisation des résidus de récoltes et des brise-vent ont par conséquent été complétés par un suivi d'exploitations intégrant l'expérience et le savoir des populations rurales (chapitre 4.3). Un résumé des conclusions tirées des textes présentés achève l'ouvrage, et délivre des recommandations pour des recherches futures et quant à l'exécution de plans de maîtrise de l'érosion par les décideurs politiques.

Ce livre est le premier d'une série de publications de recherches interdisciplinaires menées par l'Université de Hohenheim en collaboration avec des partenaires nationaux et des organisations internationales. Il est tout d'abord destiné aux chercheurs et institutions s'intéressant à l'agriculture durable en Afrique de l'Ouest sahélienne ainsi que dans d'autres régions du monde où l'érosion éolienne affecte la production agricole. Un effort particulier a cependant été réalisé pour faire en sorte que ce livre soit compréhensible pour un large lectorat incluant les experts en recherche appliquée, les vulgarisateurs, les chefs de projets et les étudiants travaillant dans le champ de l'érosion éolienne, ainsi que ceux qui 'font' les politiques nationales et internationales. Ce livre peut donner une vue d'ensemble à ceux qui sont impliqués dans la planification et l'exécution de programmes de protection des ressources naturelles, particulièrement en Afrique de l'Ouest.

B. Buerkert, B.E. Allison et M. von Oppen
Stuttgart–Hohenheim, juillet 1995

Acknowledgments

It has been a pleasure gathering scientists from different disciplines willing to contribute to this publication. We are grateful that, even though individual time and work schedules were sometimes rather tight, the authors did their best to meet the established deadlines during all stages of review and editing.

With thanks we acknowledge the constructive and valuable suggestions on earlier drafts of the papers, particularly from Prof. Dr. Dr. h.c. Horst Marschner (Part 1 to 4), Dr. Ernest Kirkby (Chapter 2.1), and Prof. Dr. Thomas Bierscheck (Chapter 4.3).

We are indebted to the Sahelian experts, Dr. John P.A. Lamers and Dr. Andreas Buerkert, for many stimulating discussions on agricultural problems in the Sahel and on wind erosion in particular, and for encouragement to proceed during difficult periods.

Our compliments to Mrs. Fadila Le Meur who did the French translations, to Mrs. Jewell Kidd for her expert assistance during the final proof-reading, and to Ms. Monika Ellinger for patiently editing the final corrections.

The publication would not have been accomplished without the general support provided by the Center for Agriculture in the Tropics and Subtropics, its staff members, and its infrastructure. We are indebted, in particular, to Dr. Jürgen Greiling for his unconditional backing throughout all stages of this project.

The Deutsche Forschungsgemeinschaft (DFG) made the research on wind erosion as well as publication of this book possible with its generous assistance.

We are thankful to the ICRISAT Sahelian Center, Niamey, Niger for hosting the Special Research Program 308 and providing logistical support.

Remerciements

Ce fut un grand plaisir que de réunir des scientifiques provenant de différentes disciplines et désireux de contribuer à cette publication. Nous sommes reconnaissants envers les auteurs qui ont fait de leur mieux pour respecter les délais à toutes les étapes de la rédaction et de l'édition, malgré leurs charges individuelles et des calendriers de travail plutôt serrés.

Nous savons gré pour les suggestions constructives et judicieuses relatives aux premières versions des textes au Prof. Dr. Dr. h.c. Horst Marschner (parties 1 à 4), au Dr. Ernest Kirkby (chapitre 2.1) et au Prof. Dr. Thomas Bierschenk (chapitre 4.3).

Nous sommes redevables aux experts du domaine sahélien, le Dr. John P.A. Lamers et le Dr. Andreas Buerkert pour de nombreuses et stimulantes discussions relatives aux problèmes agricoles au Sahel et en particulier à l'érosion éolienne, ainsi que pour leurs encouragements à agir dans des moments difficiles.

Nos compliments vont à Mme Fadila Le Meur qui a traduit en français les résumés des chapitres et la synthèse finale et Mme Jewell Kidd pour son assistance experte lors de la relecture de l'épreuve finale, ainsi qu'à Mlle Monika Ellinger pour les ultimes corrections.

Cette publication n'aurait pu être réalisée sans le soutien de l'équipe du Centre d'agriculture tropicale et subtropicale et la possibilité d'utiliser ses infrastructures. Nous avons une dette particulière envers le Dr. Jürgen Greiling pour son appui inconditionnel à toutes les étapes de ce projet.

De par sa généreuse assistance, la société allemande pour la recherche (Deutsche Forschungsgemeinschaft, DFG) a rendu possible l'édition et la publication de ce livre.

Nous sommes reconnaissants envers le Centre Sahélien de l'ICRISAT, Niamey, Niger qui a acceuilli le programme spécial de recherche 308 et a fourni une aide logistique.

List of Abbreviations

a	annum, year
Al	aluminium
ANOVA	analysis of variance
BCR	benefit/cost ratio
BSNE sampler	Big Spring number eight soil sampler
C	carbon
Ca	calcium
CEC	cation exchange capacity
CP	crude protein
CR	crop residues
cv.	cultivar
DAE	days after emergence
DAExp.	days after exposure
DAS	days after sowing
DM	dry matter
EC	electric conductivity
FAO	Food and Agriculture Organisation of the United Nations
FCFA	Franc CFA
Fig.	figure
H	distance from a windbreak in terms of windbreak height
ICRISAT	International Crops Research Institute for the Semi-Arid Tropics
IRR	internal rate of return
ITCZ	intertropical convergence zone
K	potassium
LAI	leaf area index
LDM	leaf dry matter
LP	linear programming
LSD	least significant difference
Max.	maximum
ME	metabolizable energy
Mg	magnesium
Min.	minimum
N	nitrogen
n	random sample
n.a.	not available
n.d.	not determined

B. Buerkert, B. E. Allison and M. von Oppen (eds.), Wind Erosion in Niger, xix–xx.

n.m.	not mentioned
n.s.	not significant
Na	sodium
P	phosphorus
PVINR	present value of incremental net return
S	sand
s.d.	standard deviation
SE	mean standard error
sed	standard error of the difference
sin	single grains
Sl	weakly loamy sand
TLU	tropical livestock unit
YR	soil color according to the Munsell color chart; the color term includes hue, value, and chroma.

Liste des abréviations utilisées

a	année
Al	aluminium
ANOVA	analyse de variance
BCR	ratio coût/bénéfice
BSNE sampler	collecteur de sol 'Big Spring' n° 8
C	carbone
Ca	calcium
CEC	capacité d'échange cationique
CP	protéines crues
CR	résidus de récoltes
cv.	cultivar
DAE	jours après levée
DAExp.	jours après exposition
DAS	jours après semis
DM	matière sèche
EC	conductivité électrique
FAO	Organisation des Nations Unies pour l'Alimentation et l'Agriculture
FCFA	Franc CFA
Fig.	figure
H	ratio distance au brise-vent/hauteur du brise-vent
ICRISAT	Institut International de Recherche sur les cultures des Zones Tropicales Semi-Arides
IRR	taux de rémunération interne
ITCZ	zone de convergence intertropicale
K	potassium
LAI	indice de surface foliaire
LDM	matière sèche en feuilles
LP	programmation linéaire
LSD	plus petite différence significative
Max.	maximum
ME	énergie métabolisable
Mg	magnésium
Min.	minimum
N	azote
n	échantillon aléatoire
n.a.	non disponible

B. Buerkert, B. E. Allison and M. von Oppen (eds.), Wind Erosion in Niger, xxi–xxii.

n.d.	non déterminé
n.m.	non mentionné
n.s.	non significatif
Na	sodium
P	phosphore
PVINR	valeur actualisée de l'augmentation du revenu net
S	sable
s.d.	déviation standard
SE	erreur moyenne standard
sed	erreur standard de la différence
sin	grains séparés
Sl	sable faiblement glaiseux
TLU	unité de bétail tropical
YR	couleur de sol selon Munsell; le terme 'couleur' inclue la nuance, la clarité et l'intensité.

List of Contributors

ALLISON, Bruce E.
Department of Environmental Analysis and Management, Troy State University, Troy, Alabama 36082, USA
Formerly Institute of Plant Production in the Tropics and Subtropics (380), University of Hohenheim, D-70593 Stuttgart, Germany

BAIDU-FORSON, Jojo
ICRISAT Sahelian Center, B.P. 12404, Niamey, Niger

BANZHAF, Joachim
Institut für Projektplanung, Fichtelbergstrasse 36, D-70469 Stuttgart, Germany

BATIONO, André
ICRISAT Sahelian Center/IFDC, B.P. 12404, Niamey, Niger

BLEICH, Klaus E.
Institute of Soil Science and Land Evaluation (310), University of Hohenheim, D-70593 Stuttgart, Germany

BRUENTRUP, Michael
Institute of Agricultural Economics and Social Sciences in the Tropics (490), University of Hohenheim, D-70593 Stuttgart, Germany

BUERKERT, Andreas
Institute of Plant Nutrition (330), University of Hohenheim, D-70593 Stuttgart, Germany
At present ICRISAT Sahelian Center, B.P. 12404, Niamey, Niger

BUERKERT, Barbara
Center for Agriculture in the Tropics and Subtropics (793), Special Research Programme 308, University of Hohenheim, D-70593 Stuttgart, Germany
At present ICRISAT Sahelian Center, B.P. 12404, Niamey, Niger

ELTROP, Ludger
Brünnelesberg 17, D-73760 Ostfildern, Germany

FEIL, Petra R.
Institute of Rural Sociology, Agricultural Extension, and Applied Psychology (430), University of Hohenheim, D-70593 Stuttgart, Germany

B. Buerkert, B. E. Allison and M. von Oppen (eds.), Wind Erosion in Niger, xxiii–xxiv.

HEIDHUES, Franz
Institute of Agricultural Economics and Social Sciences in the Tropics (490), University of Hohenheim, D-70593 Stuttgart, Germany

HERRMANN, Ludger
Institute of Soil Science and Land Evaluation (310), University of Hohenheim, D-70593 Stuttgart, Germany

HAMMER, Klaus
Institute of Soil Science and Land Evaluation (310), University of Hohenheim, D-70593 Stuttgart, Germany

LAMERS, John P.A.
Institute of Agricultural Economics and Social Sciences in the Tropics (490), University of Hohenheim, D-70593 Stuttgart, Germany

LEIHNER, Dietrich E.
Institute of Plant Production in the Tropics and Subtropics (380), University of Hohenheim, D-70593 Stuttgart, Germany

MARSCHNER, Horst
Institute of Plant Nutrition (330), University of Hohenheim, D-70593 Stuttgart, Germany

MICHELS, Karlheinz
ICRISAT Sahelian Center, B.P. 12404, Niamey, Niger
Formerly Institute of Plant Production in the Tropics and Subtropics (380), University of Hohenheim, D-70593 Stuttgart, Germany

OPPEN von, Matthias
Institute of Agricultural Economics and Social Sciences in the Tropics (490), University of Hohenheim, D-70593 Stuttgart, Germany

SIVAKUMAR, M.V.K.
ICRISAT Sahelian Center, B.P. 12404, Niamey, Niger

STAHR, Karl
Institute of Soil Science and Land Evaluation (310), University of Hohenheim, D-70593 Stuttgart, Germany

WILLIAMS, Timothy O.
ICRISAT Sahelian Center/ILRI, B.P. 12404, Niamey, Niger

PART ONE

Introduction

1.1
Climate of Niger

M.V.K. Sivakumar and Karlheinz Michels

Introduction

The republic of Niger (Fig. 1.1–1), covering 1 267 000 km^2, lies within the northern Tropics between 11°33' N and 23°33' N. It is bordered on the north by Algeria and Libya, on the west by Mali and Burkina Faso, on the south by Benin and Nigeria, and on the east by Chad. Over a third of the total area in Niger has inadequate rainfall for agricultural production, and agriculture is concentrated in the southern and southwestern border areas (Chapter 1.3).

Niger is one of the regions suffering most from wind erosion problems (UNEP 1992). During the rainy season from May/June to September/October, the prevailing winds blow from the southwest, disturbed by 10 to 20 strong eastern convective wind and rain storms. Storms of about 15 to 60 min duration, including gusts with maximum wind speeds up to 100 km h^{-1}, regularly precede these rainfall events, particularly during the first half of the rainy season when soils are bare (Sivakumar 1989). They transport considerable amounts of soil which bury young seedlings (Chapter 2.2) and produce impressive red dust clouds that reduce visibility, sometimes to less than 30 m (Plate 1).

Dust loads in the Sahara and the US Great Plains regions are estimated between 97 to 750 million tons per year, but most of the material is moved from one location to another at the soil surface (Fryrear 1990). Raulin (1965) noticed the loss of a thin, fertile top layer due to wind erosion in Niger. As reported in this book, wind erosion on these sandy soils caused crop losses, reduced fertility and removed fine soil fractions (Chapters 2.2, 3.3, and 3.4).

Several factors favor the occurrence of wind erosion in Niger. Increasing cultivation of marginal areas, overgrazing, the removal of trees and shrubs (Plate 2) as well as the occurrence of droughts all contribute to the loss of soil cover (Mainguet and Chemin 1977; Sivakumar 1992a). The low contents of organic matter and fine soil particles prohibit the building of non-erosive aggregates on the sandy soils. The loose sand on the surface crust is easily

B. Buerkert, B. E. Allison and M. von Oppen (eds.), Wind Erosion in Niger, 3–22.

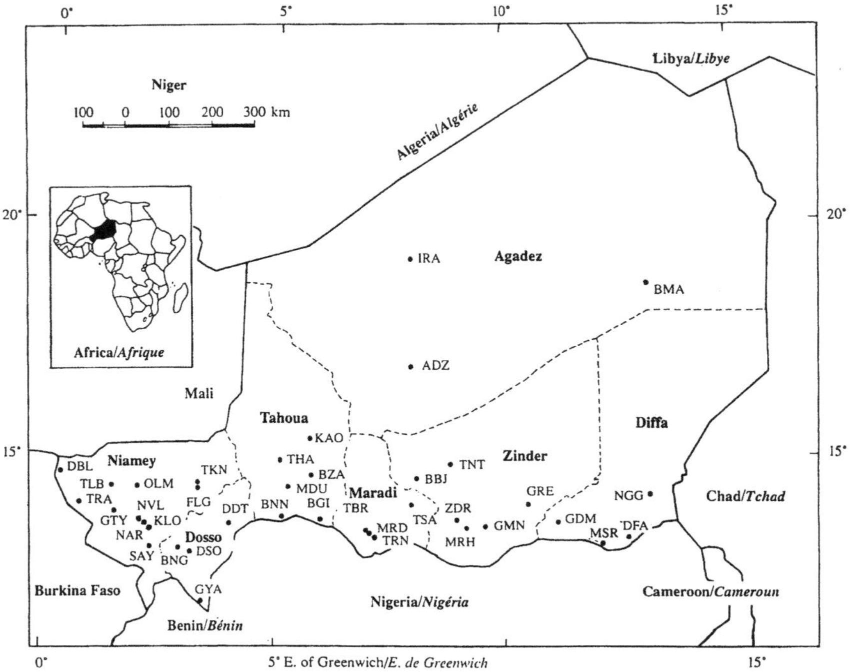

Fig. 1.1–1. Map of Niger showing the location of selected rainfall stations (for station names, see Appendix 1.1).

blown away. The susceptibility of the soils to erosion is reinforced by a reduced vegetative cover. Texture and climatic conditions favor rapid surface drying which makes it easier for erosive forces to translocate soil material. Large planting distances for millet (1 to 2 m) and slow crop establishment increase the damaging effects of wind and wind-blown soil. Unfavorable conditions in some years force farmers to replant the millet crop several times due to combined water and erosion stress (Chapter 4.3).

Any pragmatic planning for coping with wind erosion in Niger, as discussed in this book, requires not only a thorough understanding of the prevailing farming systems (Chapter 1.3 and Part 4) but also of climate. In this chapter, we review the salient features of the climate in Niger.

Physiography of Niger

In Niger, the basic physical features are massive and rugged. Low lateritic plateaus and broad interior sandy basins are rimmed by hills and limestone or sandstone bluffs. The northern Saharan part of Niger is dominated by the Aïr mountains and bordered in the northeast by the high plateaus of Tchigai and Hamada Mangueni. The mountainous areas are rugged with steep slopes, some of which are volcanic.

The region south of the 16° N latitude is about 400 000 km^2 and extends approximately 1500 km from west to east. This area of low relief plains, dissected by fossil or seasonal valleys and diversified by rocky outliers, is the region where most of the people of Niger live (Plate 3). In the western projection of the plains, the Niger River cuts through a rather even plain, mostly above 200 m in altitude. This plain has widespread surfaces of laterite.

Climatic Zones in Niger

In Niger, dry and humid tropical air converge. The area of contact between the dry tropical air and humid equatorial air is called the intertropical convergence zone, or ITCZ. The characteristic movement of the ITCZ and the disturbances associated with this phenomenon, bring summer rainfall in Niger. Winter drought is related to the presence of a subtropical anticyclone with its subsidence and divergence and the southward shifting of the ITCZ. Between 12° N and 14° N, the rainy season becomes shorter and the dry season becomes longer as one moves away from the equator. Sivakumar (1986) identified two climatic zones in the area: the southern Sahelian (growing season of 60–100 days, 400–600 mm annual rainfall) and the northern Sudanian (growing season of 100–125 days, 600–800 mm annual rainfall) zones. The typical natural vegetation is the savannah. The savannah rarely exists in a pure state and is scattered with trees and shrubs (Plate 4). At the end of the rainy season in the northern savannah, the grass dries out rapidly and yields to bare soil. When the dry season arrives the continental northeasterly wind blows; this dry wind is called the 'Harmattan' in West Africa (Plate 5).

The northern Sahelian zone marks the transition from tropical climate to desert climate. Located between 14° N and 18° N, rainfall occurs within a short period, that is a maximum of three months at its southern end and sometimes only a few weeks towards the north. The dry season which prevails for the remaining part of the year is accentuated by the continental dry winds. Vegetation in this region is distinctly xerophilous and is adapted to the dry season of more than nine months. Instead of the tall grass savannah, there is a discontinuous grassy carpet where short and prickly grasses predominate along with scattered spiny plants, small spiny- or velvety-leaved bushes, or spiny bushes.

Near 20° N, where the ITCZ disappears, the zone of continual dryness or desert begins. In the desert region, because of extreme aridity, there is little vegetation. The vegetation is adapted to very long periods of dryness; roots are large and penetrate deep, organs of evaporation are small (spines replacing leaves), grasses are hard and short.

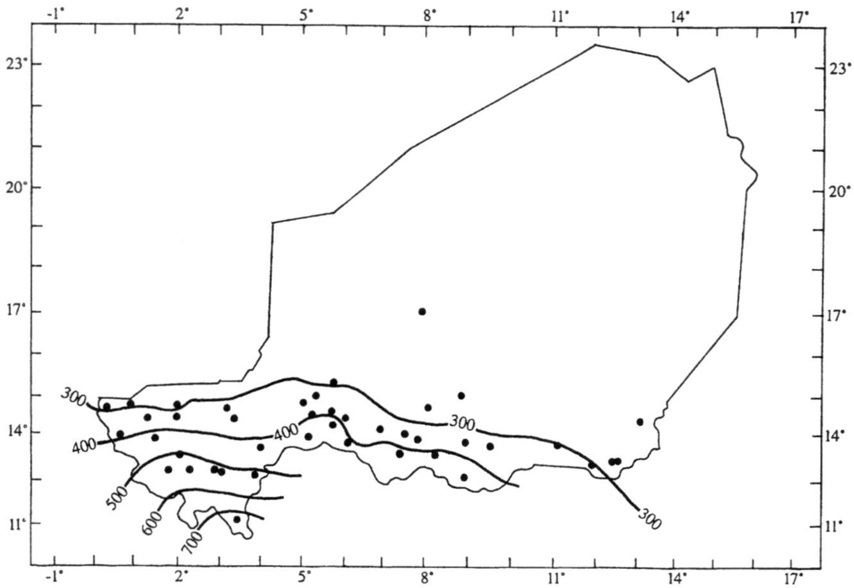

Fig. 1.1–2. Mean annual rainfall (in mm) in Niger. (*Source*: Sivakumar *et al.* 1993).

Rainfall

The list of stations used for rainfall analysis in Niger is given in Appendix 1.1 and station locations are shown in Fig. 1.1–1. The mean annual rainfall permits useful comparisons of agricultural potential. This is particularly true in the case of Niger, where seasonal patterns of rainfall are pronounced and are influenced by seasonal changes in solar radiation and pressure patterns.

Annual Rainfall Pattern

The mean annual rainfall isohyets in Niger run more or less parallel in an east-westerly direction (Fig. 1.1–2). The influence of the movement of the ITCZ, and its northern limits of activity, is immediately reflected in the latitudinal variation of rainfall. North of the 18th parallel the region is completely dry, while to the south the isohyets show a strong rainfall gradient. In the Niamey and Dosso areas, the gradient is particularly steep, reflecting large increases in the mean annual rainfall amounts as one moves southward. Mean annual rainfall at Sadoré (ICRISAT Sahelian Center) is 545 mm.

Another feature of the annual rainfall pattern is that the isohyets at the eastern end of Niger do not show the same kind of steep gradient reflected at the western end. This may be due to a greater amplitude of movement, and therefore a swifter passage of the ITCZ in the east than in the west, and to usual large-scale convergence. As Cochemé and Franquin (1967) explained, this would mean that the same annual rainfall and approximately the same length of the humid season are associated with different patterns of day length during the year, since this depends on latitude only.

Annual Rainfall Variability

Temporal variability of the annual rainfall (1931–1990) for ten locations in Niger is shown in Fig. 1.1–3. Mean annual rainfall at Agadez (Fig. 1.1–3A) is low and there is considerable inter-annual variation around the mean. However, the rainfall from 1969 onwards has been consistently below the long-term average. At Tahoua, the inter-annual variation is very large (Fig. 1.1–3B) but the rainfall variation after 1969 appears to fit the general pattern of variation. However at Filingué, located west of Tahoua, the annual rainfall shows a marked decrease from 1955 onwards (Fig. 1.1–3C). At Tillabéry, located further south, the annual rainfall was below average (Fig. 1.1–3D) for some years after 1969, but the pattern was not consistent. In contrast, at Niamey (Fig. 1.1–3E) there is no distinct decrease in rainfall after 1969 compared with the earlier years. While at Dosso, located 120 km southeast of Niamey, there is a distinct decrease in rainfall after 1969 (Fig. 1.1–3F). Similar patterns can also be seen at Birni N'Konni (Fig. 1.1–3G) and Maradi (Fig. 1.1–3H), both located east of Dosso approximately on the same latitude. However, further east at Zinder the rainfall pattern after 1969 (Fig. 1.1–3I) appears to fit the general pattern over the entire period. A similar conclusion could also be drawn about the annual rainfall variability at Gaya (Fig. 1.1–3J), which receives the highest mean annual rainfall in Niger and is located in the northern Sudanian zone.

Since a significant decline in rainfall has occurred from 1969 onwards at many locations in Niger, Sivakumar (1989) investigated if such a decline is apparent throughout Niger by calculating average rainfall for all years before and after 1969. This study used data from 1945–1969 and 1969–1990 for 25 stations in Niger. Rainfall isohyets for the two periods in Niger showed that since 1969, a clear decline has occurred in the mean annual rainfall and the isohyets have been displaced about 100 to 150 km southward. However, there were extended periods of above- or below-average rainfall for periods other than 1969–1989 at some locations; but they were not consistent for all locations. For example, rainfall was above average during 1950–1959 at Tillabéry, during 1952–1959 at Tahoua, and during 1950–1961 at Zinder; while rainfall was below average during 1937–1944 at Tahoua, during 1924–1928 at Agadez, and during 1946–1951 at Dogondoutchi.

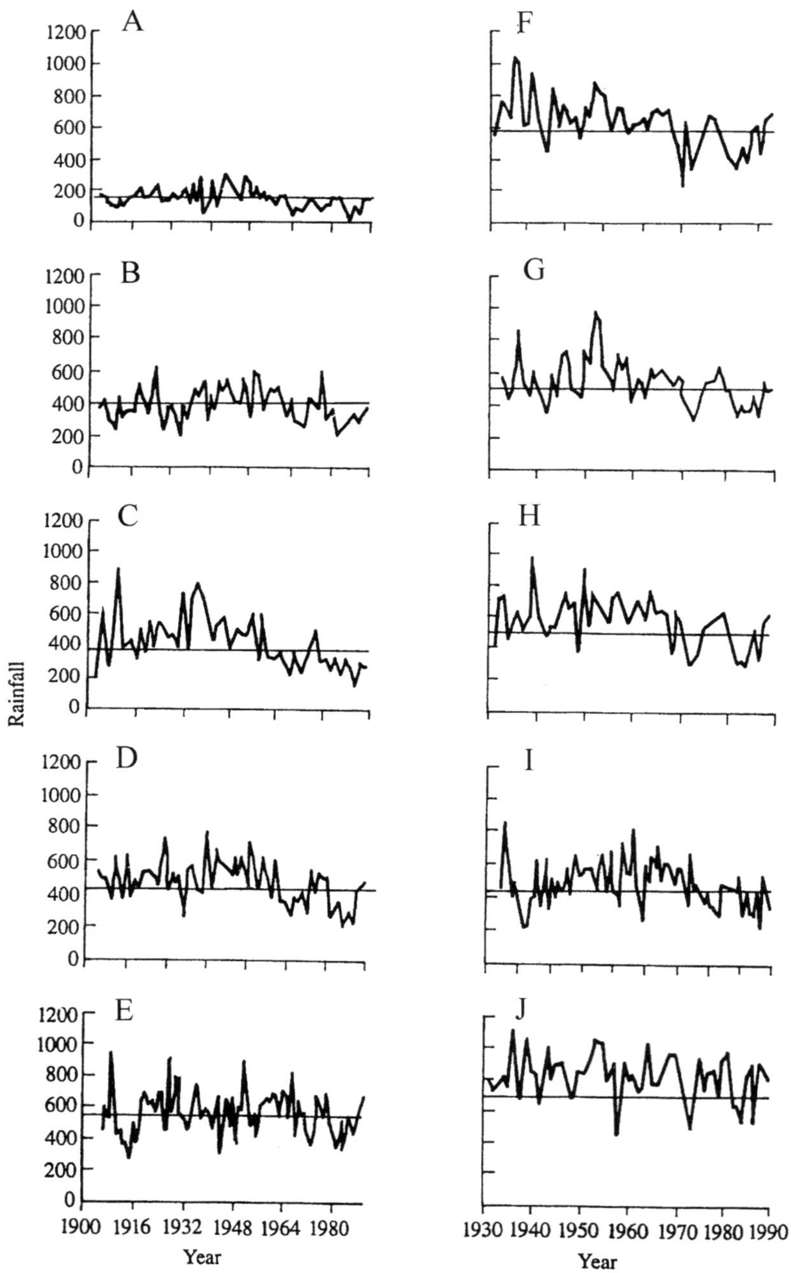

Fig. 1.1–3. Annual rainfall variability at (A) Agadez; (B) Tahoua; (C) Filingué; (D) Tillabéry; (E) Niamey; (F) Dosso; (G) Birni N'Konni; (H) Maradi; (I) Zinder, and (J) Gaya. (*Source*: Sivakumar *et al.* 1993).

Table 1.1–1. Selected statistics of monthly rainfall and rainy days at Niamey (data base 1961–1990)

Month	Rainfall				Rainy days			
	Mean (mm)	s.d.	Max. (mm)	Min. (mm)	Mean	s.d.	Max.	Min.
1	0.0	0.1	0.4	0.0	0.1	0.3	1	0
2	0.2	1.2	6.8	0.0	0.0	0.2	1	0
3	2.4	5.8	25.0	0.0	1.1	1.4	6	0
4	6.9	14.8	73.9	0.0	1.1	1.4	6	0
5	29.4	30.0	92.5	0.0	3.9	2.6	10	0
6	75.9	39.8	160.9	1.9	7.0	2.6	11	2
7	156.9	49.3	283.8	76.5	11.4	2.8	19	7
8	172.7	67.1	317.4	42.1	13.5	3.4	21	7
9	88.2	46.4	210.0	25.5	8.9	2.4	15	4
10	11.7	21.7	114.9	0.0	1.8	2.4	15	4
11	0.4	2.0	11.1	0.0	0.1	0.3	1	0
12	0.0	0.0	0.0	0.0	0.0	0.0	0	0
Total	544.8	120.9	813.0	318.9	48.3	7.4	65	30

(*Source*: Sivakumar *et al.* 1993)
s.d. = Standard deviation

Monthly Rainfall

A comprehensive picture of the degree of monthly rainfall variability within a given year and of the latitudinal variation in the distribution of monthly rainfall, emerges from histograms of monthly mean rainfall for seven locations (Gaya, Dosso, Tillabéry, Tahoua, Maradi, Zinder, and Agadez) between 11°59' N to 19°05' N latitude (Fig. 1.1–4). It is apparent that most of the rainfall is normally received over a 4- to 5-month period (May/June–September). Agadez, situated in the Sahelian desert, is an exception. Here water availability for plant growth is limited throughout most of the year, with scanty amounts during July to September. Maximum rainfall in all areas is received in August, when the ITCZ shows its full effect.

An example of monthly rainfall analysis is presented in Table 1.1–1 for Niamey. The data show that mean rainfall as well as the number of rainy days increase rapidly from April to August and then decrease to virtually zero after October. July and August are the rainiest months. Although the absolute values of standard deviations for rainfall and rainy days are the highest in July and August, as a percentage of the means they are lower in these months than in the other months during the rainy season. On average there are 48 rainy

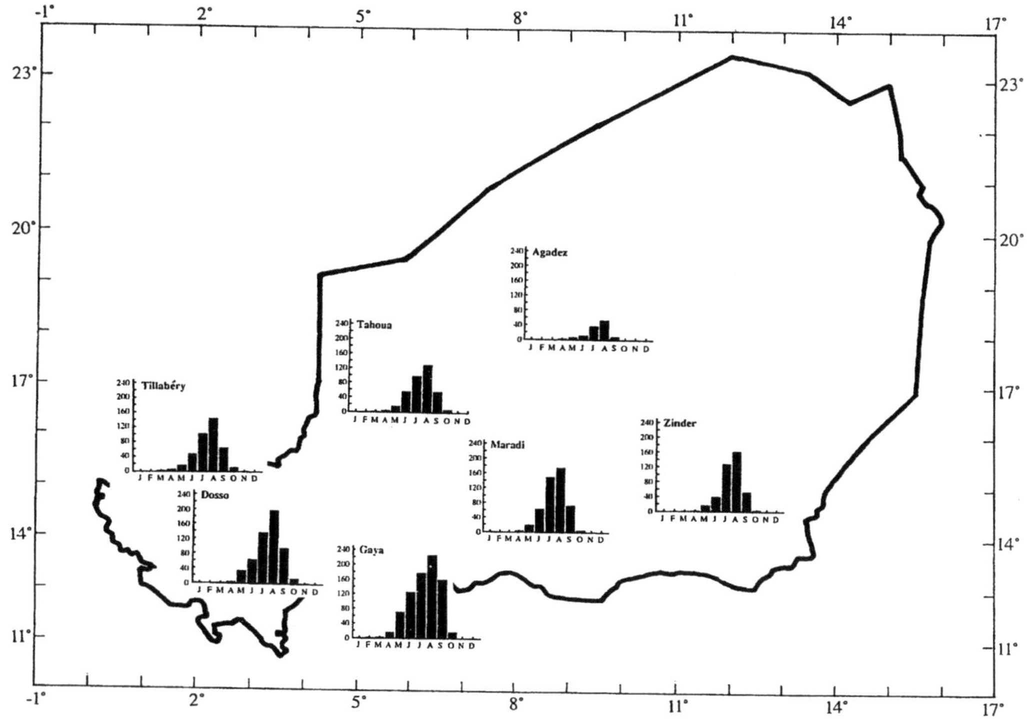

Fig. 1.1–4. Variability in monthly rainfall for selected locations in Niger. (*Source*: Sivakumar *et al.* 1993).

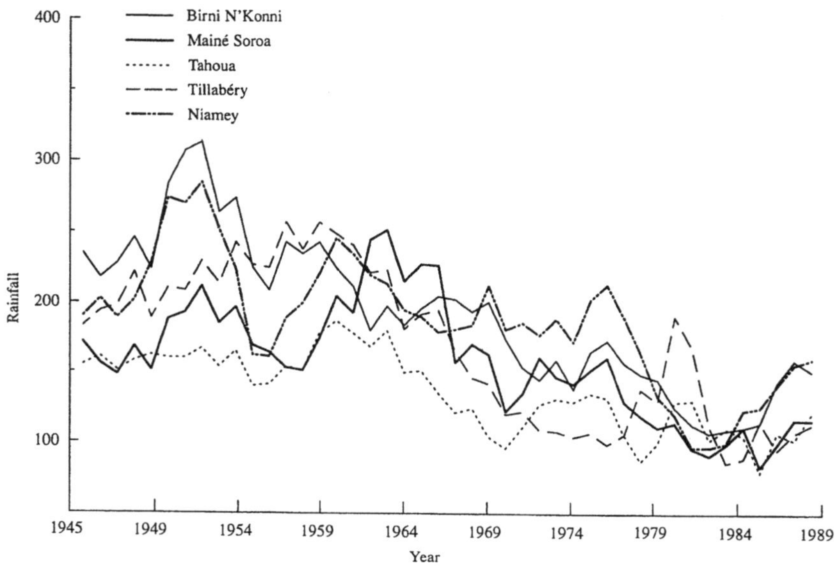

Fig. 1.1–5. Five-year moving averages of August rainfall for selected locations in Niger. (*Source*: Sivakumar *et al.* 1993).

days a year in Niamey and 45 of these are registered in the 5-month period from May to September.

In view of the importance of August rainfall, Sivakumar (1992a) plotted 5-year moving averages for 16 stations in Niger over an identical data base from 1944–1988. Data presented for seven locations show a clear decrease over the period 1965–1988 (Fig. 1.1–5).

Probability Estimates of Rainfall

Rainfall data in the form of arithmetic averages, described above for annual, monthly, and decadal scales, provide a general idea of the rainfall patterns. But averages often conceal the inter-annual fluctuations. In order to develop suitable strategies for agricultural planning and implementation, it is important to have some estimates of rainfall amounts in a given geographical region. For this purpose, Sivakumar *et al.* (1993) analyzed the decadal precipitation totals for the probability of receiving rainfall greater than several threshold values.

Figure 1.1–6 depicts the probability (percent) of receiving 10 mm or more of precipitation during the four selected decades. From 11–20 May the probability is low, ranging from less than 20% in many parts of Niger to a maximum of 69% at Gaya in southern Niger (Fig. 1.1–6A). By decade 18 (21–30 June),

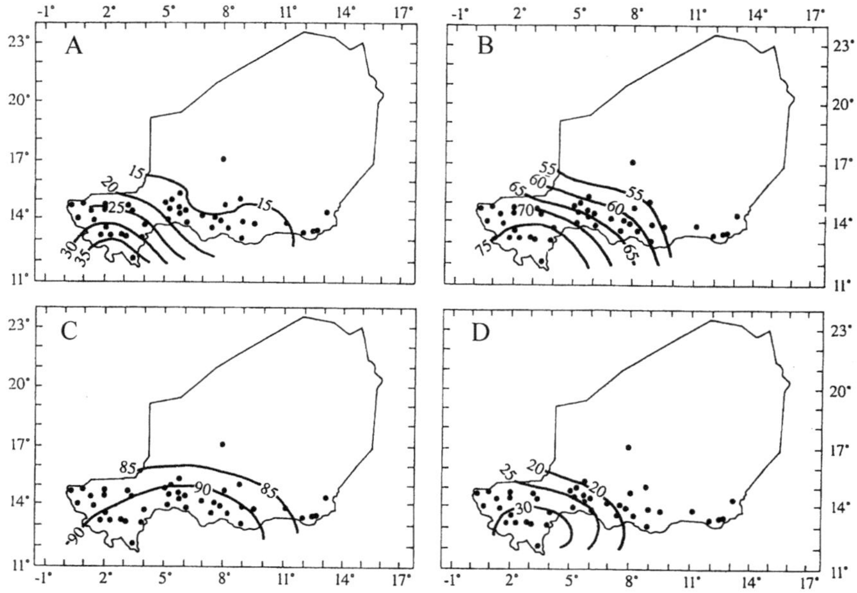

Fig. 1.1–6. Probability (in %) of receiving 10 mm or more rainfall during (A) 11–20 May, (B) 21–30 June, (C) 21–31 July, and (D) 1–10 October in Niger. (*Source*: Sivakumar *et al.* 1993).

the probabilities are appreciably higher as by this time the rains normally begin over a large part of Niger (Fig. 1.1–6B). In the Niamey and Dosso regions, rainfall probabilities exceed 70%. In the southern Tahoua and Maradi regions the probabilities are high, but over a large part of Zinder and Diffa, the probabilities are less than 55%. The highest probability, 94%, is recorded at Gaya. By late July (21-31 July), the rainfall probabilities show a much more homogenous pattern to the regions south of 14° N latitude with a probability of 90% or higher (Fig. 1.1–6C). The picture is more or less similar until decade 24 (21–31 August) by which time the ITCZ reaches its peak activity, this also happens to be the most active period of rainfall in Niger. By early October (1–10 October) when the ITCZ recedes and moves south of the equator, rainfall probabilities show a sudden decline marking the beginning of an intense dry period which extends from October to April (Fig. 1.1–6D).

Beginning and End of Rains, and Length of the Growing Season

The dates of onset and ending of rains, and the length of the growing season are important issues in planning agricultural operations. Sivakumar *et al.* (1993) computed these data for a number of stations in Niger and prepared maps of average dates of onset and ending of rains in Niger. The isolines show a large variation in the onset of rains in Niger, i.e. from 30 May in southwestern

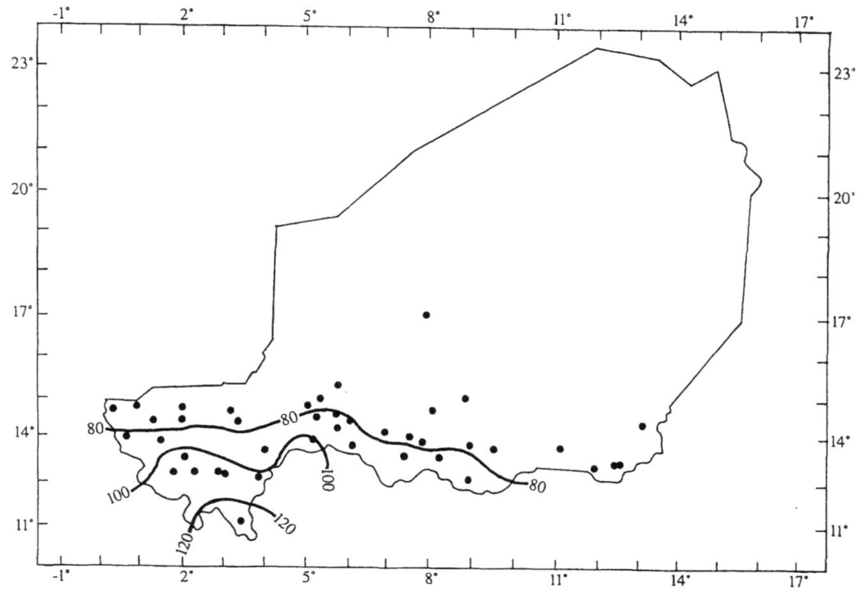

Fig. 1.1–7. Average length of growing season (in days) in Niger. (Source: Sivakumar *et al.* 1993).

Niger to 9 July in northwestern Niger. The ending of rains is comparatively rapid (Sivakumar *et al.* 1993) since the isolines show a variation of only 20 days between 7 and 27 September. Average growing season length in Niger varied from 37 days at Agadez to 138 days at Gaya. Isolines of the average length of the growing season are shown in Fig. 1.1–7. In most of the millet growing regions of Niger, the average length of the growing season varied from 80 to 110 days. In view of the temporal and spatial variability in the lengths of the growing season observed in Niger, it is important to calculate the probabilities for growing seasons of varying durations for a given date of onset of rains. Sivakumar *et al.* (1993) computed such probabilities for a number of stations in Niger.

The data for Niamey (Table 1.1–2) show that an early onset of rains offers a proportionally longer growing season while delayed onset results in a considerably shorter growing season. If the onset of rains at Niamey occurs 20 days earlier relative to the average date of onset of rains (8 June), i.e. by 19 May, there is a 99% probability that the growing season will exceed 90 days and a 49% probability of exceeding 130 days. On the other hand, if the rains are delayed until the end of June, there is only a 49% probability that the growing season will exceed 90 days. The implications of the above analysis are that crop management practices in Niger may have to be altered depending upon the onset of rains. If the rains start early, it may be safe to plant crop

Table 1.1–2. Probabilities of growing season length exceeding specified durations for variable onset of rains for Niamey, Niger

Date of onset of rains	Length of growing season (days) exceeding			
	90	110	130	150
19 May	99	88	49	11
29 May	97	70	29	2
8 June	88	49	11	0
18 June	70	29	2	0
28 June	49	11	0	0

(*Source:* Sivakumar *et al.* 1993)

cultivars recommended for median season length. If the rains are delayed until 28 June, short duration cultivars that will mature in the remaining growing season may be more productive.

Dry Spells

The analysis described so far examined mainly the rainy periods; but for some specific applications in agriculture it is equally important to consider the *rainless* periods. A simple empirical analysis of long-term rainfall data for dry spells could provide much useful information (Sivakumar 1992b).

The probabilities of a dry spell exceeding 7 days within the next 30 days are shown for four specific time periods (Fig. 1.1–8). For the period starting from 11 May, the probabilities of a 7-day dry spell exceed 95% throughout Niger, except for locations south of 14° N latitude (Fig. 1.1–8A). By 21 June, the probabilities decrease to less than 90% at most locations and to less than 70% south of 14° N (Fig. 1.1–8B). By 21 July, when the rains occur regularly throughout Niger, the probabilities of dry spells decrease to less than 70% throughout the country (Fig. 1.1–8C). But early October, when the rains recede from Niger, there is a 100% probability of dry spells of over 7 days at all locations in Niger.

Temperature

The temperature of the environment is mainly a manifestation of the radiation and energy balance. Because of uniformly high radiation levels throughout the year, Niger has a warm environment that is seasonally modified by changes in water regimes and surface conditions. Monthly average maximum and minimum air temperatures for 13 locations in Niger are given in Table 1.1–3.

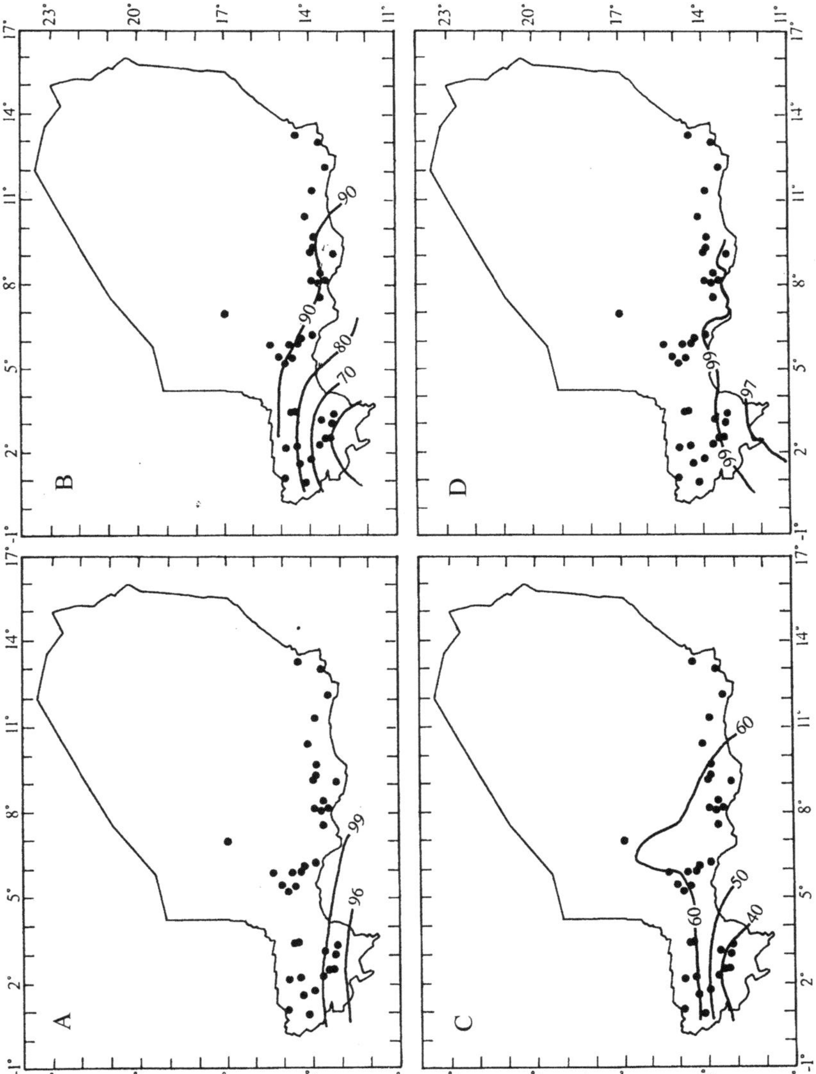

Fig. 1.1–8. Probability (in %) of maximum dry spells exceeding 7 days within the next 30 days starting from (A) 11 May; (B) 21 June; (C) 21 July; and (D) 1 October in Niger. (*Source:* Sivakumar *et al.* 1993).

Table 1.1–3. Monthly average maximum and minimum air temperature (in °C) for selected stations in Niger

Station	Jan.	Feb.	March	April	May	June	July	Aug.	Sept.	Oct.	Nov.	Dec.
Agadez	28.4	31.8	35.6	39.4	41.4	41.3	39.0	37.4	38.7	37.5	33.1	29.1
	11.3	13.5	17.8	22.4	25.5	25.6	24.5	23.7	23.8	21.2	15.8	12.4
Bilma	25.6	29.3	33.6	39.0	41.5	42.2	40.8	40.3	39.8	37.0	31.4	27.0
	8.5	10.6	15.4	20.0	23.1	24.1	25.1	25.4	22.8	18.4	12.9	9.5
Birni N'Konni	31.6	35.1	38.4	40.7	40.0	37.3	33.8	32.2	34.1	37.2	35.4	32.4
	15.6	17.8	21.7	24.9	26.6	25.3	23.6	22.6	23.0	21.7	18.6	16.2
Gaya	33.0	36.4	39.1	40.1	37.6	34.7	31.8	31.0	32.3	35.7	35.9	33.4
	18.5	21.8	25.2	27.0	26.0	24.3	23.0	22.5	22.3	22.8	20.4	19.0
Goure	29.6	33.2	36.3	40.0	41.3	39.7	35.8	34.0	36.0	37.4	34.0	30.6
	13.2	15.4	19.3	23.3	24.9	24.7	23.3	21.9	22.2	30.1	17.2	14.2
Magaria	30.7	33.9	37.6	40.5	40.6	38.1	34.1	32.7	35.2	37.5	35.0	31.4
	11.2	12.6	16.7	21.6	23.7	23.9	21.8	21.4	21.0	18.4	14.5	11.5
Maine Soroa	30.5	33.3	37.5	40.5	40.9	38.8	34.8	32.8	35.5	37.8	34.4	31.2
	12.7	15.2	19.8	23.4	25.2	24.8	23.3	22.2	22.3	21.0	16.9	13.3
Maradi	30.7	34.0	37.5	39.9	39.4	36.8	33.0	31.2	33.3	36.2	34.3	31.3
	12.8	15.2	19.7	23.5	25.3	24.5	22.8	21.8	22.0	19.9	16.1	13.2
N'Guigmi	32.9	35.9	39.0	40.9	39.9	36.9	33.7	32.1	33.9	37.3	36.4	33.4
	12.7	15.0	19.4	22.4	24.3	24.6	24.3	23.5	23.2	21.0	16.9	13.3
Niamey Ville	32.9	35.9	39.0	40.9	39.9	36.9	33.7	32.1	33.9	37.3	36.4	33.4
	15.9	18.7	22.8	26.3	27.4	25.4	23.6	22.9	23.3	23.8	19.3	16.4
Tahoua	31.5	34.6	38.2	40.8	40.8	38.5	35.0	33.1	35.4	38.0	35.3	32.1
	15.1	17.6	21.9	25.3	26.9	25.7	23.8	23.1	23.4	23.0	19.2	16.2
Tillabéry	32.7	35.8	39.1	41.4	41.0	38.4	35.3	33.2	35.1	38.3	36.8	33.4
	16.8	19.3	22.6	25.5	27.5	26.2	24.6	23.8	24.1	23.7	19.9	17.3
Zinder	29.8	33.0	36.9	39.7	40.0	37.9	33.9	32.1	34.7	37.0	34.0	30.5
	14.6	17.2	21.4	24.9	26.1	25.2	23.3	22.7	22.9	22.5	18.7	15.4

(*Source:* Sivakumar *et al.* 1993)
Figures in the top row for each station indicate the maximum air temperature and those in the bottom row the minimum temperature

During the dry season at Agadez and Bilma, located north of 16° N latitude, the mean maximum air temperatures reach up to 42°C. During the cool season from November to February, mean maximum air temperatures are below 33°C and the minimum air temperatures are below 15°C. The minimum air temperatures increase progressively to 25°C during May and vary between 18 and 25°C until the cool season begins in November.

As one moves south from 16° N latitude towards the Sahelian zone, differences between the dry season and rainy season, in both the mean maximum and minimum air temperatures, increase because temperature changes with rainfall. During the dry season, between December and May, the mean maximum temperatures vary between 30 and 41°C for stations such as Niamey, while the mean minimum temperatures range from 11 to 27°C. During the rainy season, the diurnal range in temperature decreases at Niamey but is still larger than at locations situated further south. During this period, the mean maximum temperatures range between 31 and 39°C, while the mean minimum temperatures vary between 21 and 26°C.

In the Sahelian zone, the maximum soil temperature in the upper soil horizons is of special importance for crop production at the beginning of the cropping season. Due to high solar radiation, the peak temperature can exceed 50°C at 0 to 10 mm depth, thus affecting seedling emergence and the growth of seedlings that have been covered by wind-blown soil (Chapters 2.2 and 3.3).

At Gaya (Fig. 1.1–1), the annual air temperature is less variable than in the southern and northern Sahelian zone. During the dry season, the maximum temperatures vary between 33 and 40°C, and the minimum temperatures between 18 and 27°C. During the rainy season the diurnal range in temperature is reduced and varies from 8 to 13 °C with the maximum temperatures between 31 to 38°C and the minimum temperatures from 22 to 26°C.

Potential Evapotranspiration

The water requirements of a crop at two different locations could vary, even though the rainfall amount and its distribution are similar. The atmospheric demand for water between the two locations could be quite different. The potential evapotranspiration is an indication of the potential demand for water at a given location. Monthly potential evapotranspiration rates for 13 stations in Niger were computed from maximum and minimum air temperatures, vapor pressure deficit, sunshine hours, and wind speed. The method of computation was similar to that described by Cochemé and Franquin (1967). The computed values of average monthly potential evapotranspiration for the 13 stations are shown in Table 1.1–4.

During the cooler months of November to February, the potential evapotranspiration values in northern Niger are lower than those elsewhere in Niger. Here, with increasing air temperatures and solar radiation during the rainy season, the monthly potential evapotranspiration rates in the Sahelian zone increase faster and are the highest in the country. At Gaya, in the Sudanian zone, the potential evapotranspiration rates during June to October are the lowest in the country.

Table 1.1–4. Monthly variation in the potential evapotranspiration (in mm) at selected locations in Niger

Station	Jan.	Feb.	March	April	May	June	July	Aug.	Sept.	Oct.	Nov.	Dec.
Agadez	176.2	194.3	239.9	251.2	254.6	236.9	231.5	228.7	220.6	222.0	184.1	177.0
Bilma	131.3	148.7	193.9	202.0	209.9	213.0	215.3	213.7	194.4	183.3	118.3	118.2
Birni N'Konni	151.4	161.5	200.9	199.6	215.4	193.7	178.5	166.5	165.9	171.2	148.4	139.8
Gaya	181.3	193.1	218.4	210.2	194.1	172.2	147.8	140.9	145.7	165.7	163.6	168.8
Goure	199.5	214.9	242.0	261.9	276.3	260.8	232.9	224.4	205.8	234.1	185.8	186.9
Magaria	126.6	144.3	178.7	198.0	225.5	211.8	183.6	173.9	173.9	177.2	139.0	121.1
Maine Soroa	155.8	171.2	210.5	221.7	231.7	216.4	193.3	185.8	172.2	190.8	161.0	144.3
Maradi	179.3	192.0	225.7	229.6	232.5	203.2	178.4	168.6	165.2	188.9	178.2	169.2
N'Guigmi	149.0	159.0	205.4	218.3	215.5	208.8	198.1	189.7	180.4	190.9	158.7	143.0
Niamey Ville	173.3	186.5	216.3	224.2	238.4	200.5	186.5	173.8	176.2	187.2	167.2	163.5
Tahoua	178.2	191.8	210.3	229.0	244.4	216.7	208.4	201.5	191.7	202.8	188.8	171.6
Tillabéry	145.2	158.0	192.3	199.6	220.6	195.4	185.3	175.5	171.0	173.6	145.8	133.5
Zinder	154.7	164.2	202.8	211.6	224.7	210.1	191.1	183.7	176.3	181.4	168.4	152.1

(*Source:* Sivakumar *et al.* 1993)

Conclusions

Niger is characterized by distinct climatic zones from north to south. Agricultural and pastoral activities take place in the southern Sahelian and the northern Sudanian zones where the rainy season lasts from May/June to September. Rainfall isohyets run parallel in an eastwesterly direction. The rainy season becomes shorter and rainfall decreases towards the north. From 1969 onwards, rainfall has been consistently below the long-term average at many locations. Large parts of Niger are situated in the northern Sahelian zone and in the desert region, both with very little rainfall and vegetation. Wind erosion occurs during the rainy season, mostly preceding rainfall events. In the dry season, the 'Harmattan' winds transport dust from the Sahara desert. Soil temperatures at the beginning of the cropping season may exceed 50°C in the top layer, and thus, affect seedling emergence and the recovery of seedlings covered by wind-blown soil.

Abstract

The frequency of wind erosion events in Niger in recent years makes it pertinent to examine the important features of climate in Niger. Rainfall patterns in Niger showed a significant decline in the annual rainfall after 1960. Temporal variability of rainfall in Niger on the annual and monthly scale is large. Maps showing the probabilities of rainfall and the dry spells for selected periods during the rainy season are presented. The onset of rains and the probabilities of growing season lengths of different duration for variable onset of rains are analyzed. Maximum and minimum temperatures and potential evapotranspiration for several locations in Niger show large differences from north to south.

Résumé

Le climat au Niger

La fréquence de plus en plus importante des phénomènes d'érosion éolienne au cours de ces dernières années au Niger fait de l'étude des principales caractéristiques du climat et des systèmes de production un sujet pertinent. Les schémas pluviométriques au Niger montrent une importante chute dans la pluviométrie annuelle depuis 1960. On remarque une forte variabilité de la pluviométrie dans le temps aussi bien à l'échelle annuelle que mensuelle. Ce document présente des cartes de probabilités concernant les pluies et les périodes sèches à des moments déterminés au cours de la saison des pluies. En fonction des différents moments possibles de l'installation des pluies, les durées probables des saisons des cultures sont analysées. Les températures maximales et minimales ainsi que l'évapo-transpiration potentielle à plusieurs endroits du pays montrent de fortes variations du nord au sud.

References

Cochemé, J. and P. Franquin. 1967. An agroclimatological survey of a semi-arid area in Africa south of the Sahara. Technical Note No. 86, World Meteorological Organization, Geneva, Switzerland.

Fryrear, D.W. 1990. Wind erosion: Mechanics, prediction, and control. *Advances in Soil Science* 13: 187–199.

Mainguet, M. and M.C. Chemin. 1977. Les marques de l'érosion éolienne dans le Sahel du Niger d'après les images satellites et les photographies aériennes. pp. 139–149. *In* Colloque Pédologie Télédétection, AISS, Rome, Italy.

Raulin, H. 1965. Techniques et bases socio-économiques des sociétés rurales nigériennes. Etudes nigériennes 12, IFAN-CNRS, Niamey, Niger.

Sivakumar, M.V.K. 1986. Soil-climatic zonation for West African semi-arid Tropics: Implications for millet improvement. Presented at the Regional Pearl Millet Workshop, Niamey, Niger, 6–12 Sept. 1986.

Sivakumar, M.V.K. 1989. Agroclimatic aspects of rainfed agriculture in the Sudano-Sahelian zone. pp. 17–38. *In* Soil, crop, and water management systems for rainfed agriculture in the Sudano-Sahelian zone. Proceedings International Workshop, ICRISAT, Niamey, Niger, 7–11 Jan. 1987, ICRISAT, Patancheru, India.

Sivakumar, M.V.K. 1992a. Climate change and implications for agriculture in Niger. *Climatic Change* 20: 297–312.

Sivakumar, M.V.K. 1992b. Empirical analysis of dry spells for agricultural applications in West Africa. *Journal of Climatology* 5: 532–539.

Sivakumar, M.V.K., A. Maidoukia, and R.D. Stern. 1993. Agroclimatology of West Africa: Niger. Second edition. Information Bulletin No. 5, ICRISAT, Patancheru, India / Direction de la Météorologie Nationale du Niger, Niamey, Niger. 116 pp.

UNEP. 1992. Atlas of desertification. United Nations Environment Programme (ed.) Edward Arnold, Dunton Green, U.K.

Appendix 1.1. List of the rainfall stations used for the analysis

S.No. Station	Code	Latitude (N)		Longitude (E)		Altitude	Data base	Missing
		degree	min	degree	min	(m)	(in 1961–1990)	years
1. Agadez	ADZ	16	58	7	59	498	1921–90	84
2. Ayorou	AYR	14	44	0	55	223	1954–90	
3. Bambeye	BBY	14	43	5	5	365	1959–90	87, 89
4. Bangui	BGI	13	43	6	6	350	1958–90	
5. Belbedji	BBJ	14	39	8	4	400	1959–90	84
6. Birni N'Gaoure	BNG	13	5	2	54	188	1953–90	80–82
7. Birni N'Konni	BNN	13	48	5	15	272	1933–90	
8. Bouza	BZA	14	25	6	3	300	1954–90	
9. Chetimari	CMR	13	18	12	30	320	1959–90	63, 76, 82
10. Diffa	DFA	13	19	12	37	255	1951–90	79–81
11. Dogondoutchi	DDT	13	38	4	0	230	1923–90	78–80
12. Dolbel	DBL	14	37	0	17	300	1959–90	
13. Dosso	DSO	13	1	3	11	218	1931–90	73, 78, 79, 81
14. Filingué	FLG	14	21	3	19	300	1931–90	68
15. Garhanga	GHG	14	33	5	46	360	1959–90	65, 83, 84
16. Gaya	GYA	11	59	3	30	160	1931–90	
17. Gotheye	GTY	13	49	1	35	220	1954–90	83, 85, 86
18. Goudoumaria	GDM	13	43	11	10	305	1952–90	67, 86
19. Guidimouni	GMN	13	41	9	31	370	1954–90	64, 88
20. Illela	ILA	14	28	5	15	320	1954–90	80
21. Kanambakanchi	KBK	13	52	7	49	350	1959–90	
22. Kao	KAO	15	14	5	45	400	1959–90	77, 84
23. Korgoum	KGM	13	27	8	15	370	1959–90	65, 70, 73
24. Kornaka	KNK	14	6	6	54	420	1959–90	63, 76
25. Magaria	MGI	12	59	8	56	360	1938–90	64, 66–69
26. Maine Soroa	MSR	13	14	11	59	338	1936–90	
27. Mangaize	MGZ	14	42	2	1	250	1959–90	62, 64, 76, 85
28. Maradi	MRD	13	28	7	25	368	1931–90	75–78
29. Mayahi	MYH	13	57	7	40	400	1959–90	65, 76, 77, 85
30. N'Guigmi	NGG	14	15	13	7	286	1921–90	

Appendix 1.1. Continued

S.No. Station	Code	Latitude (N)		Longitude (E)		Altitude	Data base	Missing
		degree	min	degree	min	(m)	(in 1961–1990)	years
31. Niamey Ville	NVL	13	30	2	8	216	1905–90	
32. Ouallam	OLM	14	19	2	5	250	1947–90	81
33. Say	SAY	13	6	2	21	200	1921–90	
34. Tahoua	THA	14	54	5	18	385	1959–90	85
35. Tama	TMA	14	16	5	46	350	1936–90	83, 84
36. Tanout	TNT	14	58	8	53	400	1938–90	66, 68, 70, 84
37. Tera	TRA	14	0	0	45	300	1938–90	80–81
38. Tibiri	TBR	13	6	4	0	220	1959–90	76–78, 84
39. Tillabéry	TLB	14	12	1	27	209	1923–90	
40. Torodi	TRD	13	7	1	48	215	1962–90	66, 81–84
41. Toukounous	TKN	14	30	3	17	290	1956–90	78, 81, 84, 85
42. Zinder	ZDR	13	48	8	59	451	1905–90	75–78

(*Source*: Sivakumar *et al.* 1993)

1.2
Soils of Western Niger

Klaus E. Bleich and Rainer Hammer

Land Forming Processes

Nearly all soils in western Niger, as in many other regions of West Africa, have been formed from sand. The soil associations originate from three land-forming processes that took place under quite different climatic conditions, but in similar relief situations.

The Formation of Laterite

According to Greigert (1966), the formation of laterite took place in three phases: (i) A deep weathering occurred under a moist tropical climate in the Mesozoic era. This resulted in the formation of kaolinitic clay, coarse quartz-sand, gravel, and iron oxides. (ii) These were redistributed and deposited in large depressions during the Tertiary period as 'Continental terminal'. (iii) The last phase of laterite formation was the development of an iron oxide crust. This took place in erosion and deposition areas. In erosion areas, the iron oxide crust cemented the soil surface or the topsoil, whereas in deposition areas crusts were also formed in soil layers rich in sand (Greigert 1966). The second and third phases took place under a more semi-arid climate; the latter is ongoing as debris from the crust accumulates on footplains and is re-cemented.

The Dissection of Laterite Shields

The dissection of laterite shields was caused by a powerful river system during the Pleistocene epoch. The weather was moister and somewhat cooler than current weather patterns. Large steps and inselbergs along the ancient river valleys (dallols) and the Niger River, were subsequently formed.

B. Buerkert, B. E. Allison and M. von Oppen (eds.), Wind Erosion in Niger, 23–32.

Spreading of Dune Systems

Dune systems developed in a southwesterly direction in a climate that was quite dry. These climatic conditions occurred during the late Pleistocene epoch (*Erg ancient*, > 40 000 years) and more recent time periods (*Erg recent*, 16 000 to 20 000 years). There were intensive periods of weathering between the development of the two dune systems and afterwards. In northern Niger, the development of a third dune system, which began 2500 to 3000 years ago, can be observed (Völkel 1989).

The older dune system covers an extensive area and forms the surface in the southern part of western Niger, for example at Tara or Sadoré (Plate 6). In a southwesterly direction, the older dunes have been flattened to a greater degree by water erosion and replaced by sand-washed plains.

The younger dune system covers large areas between Boubon, Maradi, and Tahoua. Its southern border is between 14° N and 13° N latitude, where the dunes are oriented along the Niger River. The younger dunes are steeper than the older dunes and were partly derived from topsoil material of the older dune system. Sand movement on a large scale is now restricted to areas where the vegetation cover has been damaged and 'step-stress' by livestock occurs. This can be observed north of Ouallam.

Soil Distribution

The soil distribution in western Niger is mainly the result of soil erosion and deposition. Wind and water erosion have, at some sites, removed the soil down to the metamorphic rocks. This has resulted in the retranslocation of dune sand and laterite material. Subsequently, deposits and re-deposits (in northwestern Niger) of dune sand were formed. Slope deposits of debris from iron oxide crusts and mixed sediments of kaolonitic clay were cemented with quartz-sand. Alluvial material was deposited in periodically flooded channels, on river banks, and in depressions. The alluvial deposits have mostly derived from the sandy soils of the dunes and from the iron oxide crusts. In dallols, the soil distribution was also affected by a high groundwater level and high salt contents.

In the older dune system, the dominating soil forming process was clay movement in the profile that caused reddish brown (5 YR) topsoils. The fine sandy soils have a single grain structure with 95% being larger than 63 μm (Chapter 3.4), and have weakly developed A horizons. The red (2.5 YR) subsoils are well-developed (> 1 m depth) and contain a few percent of kaolinitic clay. The experimental fields of the ICRISAT Sahelian Center, on which most of the research presented in the following chapters was conducted, are also located within the older dune system. In Fig. 1.2–1, profile descriptions of a typical catena in the area of Sadoré are given. The soils are mainly Luvic

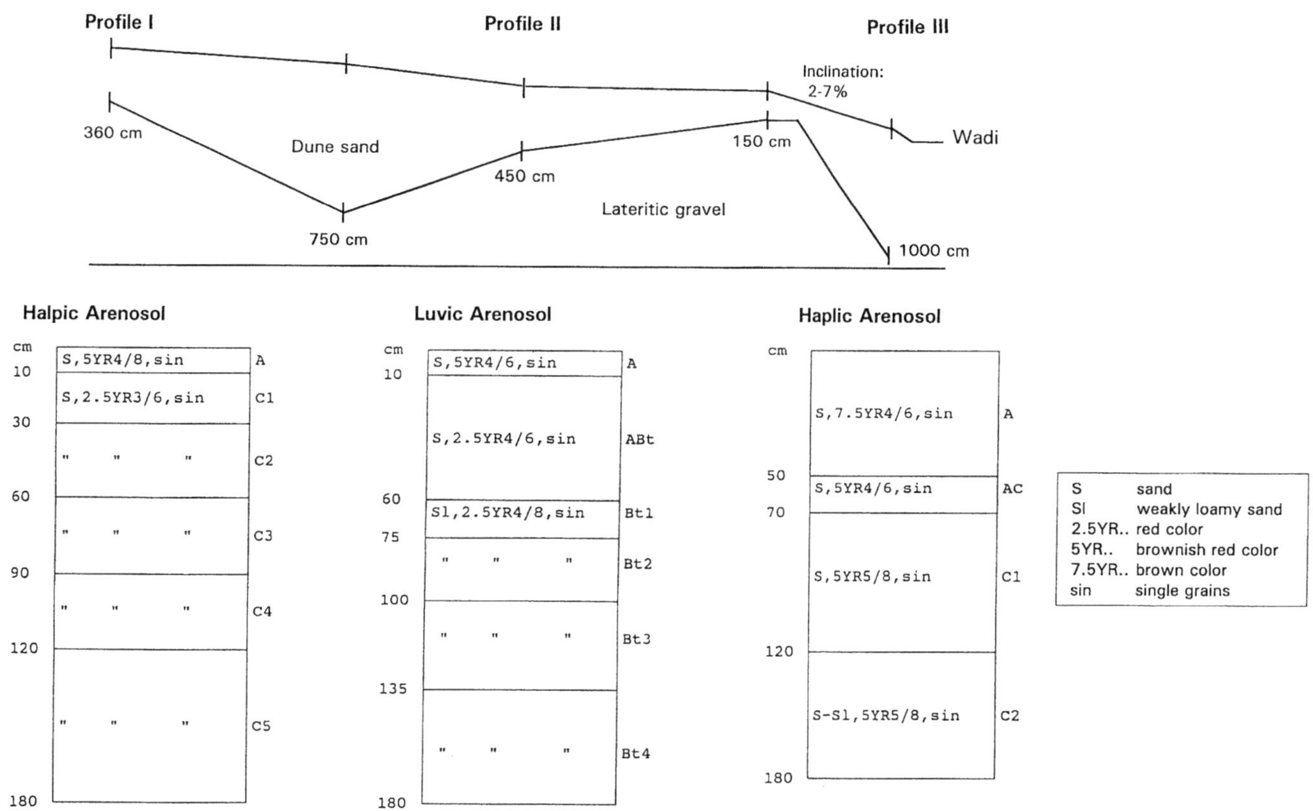

Fig. 1.2–1. Schematic presentation of a dune catena over lateritic gravel near Sadoré (western Niger), and descriptions for selected soil profiles. (*Source*: Hammer 1994, modified).

Arenosols and Halpic Arenosols according to the FAO soil classification (FAO 1988).

Within the catena, the A horizons are moderately darker and often thinner in the upper slope positions (Fig. 1.2–1). The translocation of humus particles by water erosion has increased the content of organic matter in lower slope positions (Hammer 1994). Consequently, the pH, the base saturation, and the SiO_2/R_2O_3 ratio have decreased in lower slope positions (Table 1.2–1). In general, the sandy soils have low concentrations of available nutrients. Thus, nutrient deficiencies, especially of phosphorus (P), are important constraints to crop production in the region (Bationo *et al.* 1990).

In the lower slope positions of laterite steps, mixed dune sand deposits can be found (Ferralic Arenosols). They contain more kaolinitic clay, have a fragile structure and are more sensitive to water than to wind erosion. In upper slope positions, the soils are Haplic Ferralsols and on the plateaus, Dystric Leptosols. Within the catena, there is a higher total P concentration in the plateau soils compared to the soils downslope (Table 1.2–2). In contrast, P_{Bray} tends to be lower in the plateau soils.

On the lateritic plateaus, discontinuous sand cover formed the *brousse tigrée,* where strips of shrub vegetation alternate with strips of bare laterite (Plate 7). Within the vegetated strips, a selective accumulation of dust can be observed.

The younger dune system in northwestern Niger is characterized by Ferralic and Luvic Arenosols. The Arenosols are moderately developed and have brown (7.5 YR) topsoils and reddish brown (5 YR) subsoils. There is a fine sandy texture throughout the soil profile. Compared to the soils of the older dune system, there are decreased concentrations of aluminium (Al) and increased concentrations of exchangeable bases in these soils (Hammer 1994).

Chromic Cambisols with remarkably higher loam contents and more intensive coloring of the subsoil are found at the foot of the slopes. In northwestern Niger, Eutric Regosols were developed either from unweathered sands in partly eroded sand depressions or from weathered metamorphic bedrocks. Lixisols are found on terraces with sandy topsoils above weathered waterlogged subsoils derived from crystalline rocks. Slightly weathered outcrops, rich in feldspars, are responsible for the occurrence of Calcaric Regosols.

On soils from dune sand, crust formation is a well-known phenomenon. Hoogmoed and Stroosnijder (1984) distinguished two types of crusts found on sandy Sahelian soils. The first crust type is thin (< 0.1 mm thick), but it can seriously impede infiltration (Table 1.2–3). It occurs on sandy soils with only a small percentage of clay or silt. The impact of raindrops separates clay and silt particles which are then translocated by the infiltrating water and then block the soil pores (Chen *et al.* 1980). Deposited 'Harmattan' dust (Chapter 2.1) can intensify this process. The second crust type forms crusts which are easily visible (thickness > 1 mm). They are frequently formed on

Table 1.2–1. Chemical properties of two Halpic Arenosols from a dune catena near Sadoré (Profile I and III; Fig. 1.2–1)

	C_{total}	N_{total}	C/N	Conductivity	pH	Al^{3+}	SB[a]	CEC[b]	BS[c]	Si	Al	Fe	$SiO_2/$	Ca	Mg	K	Na	Mn	P	P_{Bray}
	— (‰) —			(μS cm^{-1})	($CaCl_2$)	— ($cmol_c$ kg^{-1}) —			(%)	— (‰) —			R_2O_3	——— (‰) ———					– (mg kg^{-1}) –	
Profile I																				
A	2.3	0.28	8.2	19	4.7	0.17	1.05	2.07	33	360	28.0	16.1	11.7	0.8	1.1	0.5	2.0	80	220	5.1
C2	1.2	0.10	12.0	7	3.6	0.13	0.32	1.37	38	390	23.4	13.5	25.1	0.8	1.0	0.4	1.3	90	180	4.5
C3	0.6	0.08	20.0	11	4.3	0.04	0.41	1.43	33	408	16.8	11.7	35.1	0.8	0.9	0.1	0.2	90	140	4.3
C4	1.2	n.d.	n.d.	9	4.9	0.04	0.55	1.90	39	400	18.7	12.5	31.2	0.8	0.9	0.2	1.2	70	160	3.8
C5	1.2	n.d.	n.d.	10	5.2	0.01	0.57	1.45	28	402	15.9	11.2	36.4	0.8	0.8	0.2	0.7	80	160	n.d.
Profile III																				
A	2.3	0.10	23.0	7	3.5	0.41	0.05	1.15	4	375	13.4	9.2	40.5	0.4	0.3	0.6	1.2	60	160	3.3
AC	1.6	0.08	20.0	9	3.6	0.29	0.10	1.53	7	374	17.2	11.3	31.9	0.4	0.3	0.8	0.7	70	150	3.4
C1	1.0	0.07	14.3	6	3.5	0.24	0.33	1.50	22	368	20.0	11.3	27.7	0.4	0.3	0.8	1.0	60	150	3.5
C2	n.d.	n.d.	n.d.	5	4.2	0.03	0.37	1.48	25	n.d.	n.d.	n.d.	n.d.	n.d.	n.d.	n.d.	n.d.	n.d.	n.d.	n.d.

(*Source*: Hammer 1994, summarized)
n.d. = not determined
[a] Sum of bases
[b] Cation exchange capacity
[c] Base saturation

Table 1.2–2. Total phosphorus (P_t), available phosphorus (P_{Bray}), and total carbon (C_t) in soils from laterite plateaus, foot of slopes, and depressions at three sites in southwestern Niger

Profile	Horizon	P_t (mg kg^{-1})	P_{Bray} (mg kg^{-1})	C_t (g kg^{-1})
Gaya I (250 km SE)[a]				
Plateau	AhBsc	2110	3.2	7.7
Foot of slope	A1	540	3.8	2.5
	A2	530	8.0	3.0
	2Bs1	460	2.9	2.4
Depression	A1	230	4.9	3.7
	A2	220	3.7	3.8
	2Bs1	350	1.9	2.3
Koigorou (112 km ESE)				
Plateau	A	360	8.0	6.9
Depression	1A	130	9.3	4.8
	2A	150	1.3	2.4
	2C	70	0.0	n.d.
Kobadié (60 km SSW)				
Plateau	A	340	12.3	5.5
	2Bsc1	340	4.9	4.8
	2Bsc2	330	3.5	4.1
Foot of slope	crust[b]	160	4.9	1.5
	A	170	2.5	1.2
	C	160	1.5	0.9
Depression	A1	200	8.2	7.6
	A2	180	4.2	3.0
	BW1	200	4.9	2.6

(*Source*: Hammer 1994, summarized)
n.d. = not determined
[a] Distance and compass direction from Niamey
[b] Surface crust

Table 1.2–3. Range of infiltration rates of different soils in southwestern Niger

Soil type	Infiltration $mm\ h^{-1}$
Arenosol, non-crusted (dune)	360–600
Arenosol, crusted (dune)	2
Leptosol (laterite plateau)	8–70
Gleysol (valley bottom)	8–30
Cambic Lixisol (terrace)	5–120

(*Source*: Hammer 1994)

soils that have poor aggregate stability. When the soils are moist, the crust will reduce infiltration and when dry, will form a barrier against seedling emergence (Hoogmoed and Stroosnijder 1984). Both types of hardened bare surfaces are polished by wind-transported sand grains during the convective storms.

The deeply developed soils from dune sand cover more than 50% of the surface in western Niger. While nearly 30% of the soils belong to the laterite plateaus. A comparison of the plateau soils reveals that the soils in areas with higher rainfall in southwestern Niger are more affected by water erosion than soils in northwestern Niger. From Gaya to Kobadié, an increase in pH (Hammer 1994) and in concentrations of exchangeable bases, especially of calcium (Ca) and magnesium (Mg), as well as a decrease in concentrations of Al can be observed (Table 1.2–4). The organic matter contents increase from the south to the north (Hammer 1994; Table 1.2–5).

Loamy to clayey Gleysols occur on the lowest terraces of the Niger River valley and in the ancient channels of the dallols. They cover approximately 20% of the dallol surface. Similar to the neighboring dune soils, they are Dystric in the south and Eutric in the north. However, the Al/Fe ratio is increased compared to the dune soils and the concentrations of exchangeable bases are markedly decreased. Thus in Gleysols, a high risk of Al toxicity must be taken into consideration (Hammer 1994). In clayey depressions with minimal runoff, Vertisols have developed from material rich in bases. Orthic Solonchaks have formed in soils with high concentrations of Ca, Mg, and sodium.

The overview of the soils of western Niger and their physical and chemical properties shows that dune sands are the predominant parent soil material. The susceptibility to wind erosion of the soils formed from dune sand increases from southwestern to northwestern Niger.

Table 1.2–4. Exchangeable cations in soils (in $mmol_c\ kg^{-1}$) from laterite plateaus, foot of slopes, and depressions at two sites in southwestern Niger

Profile	Horizon	Ca^{2+}	Mg^{2+}	K^{+}	Na^{+}	Sum of bases	Al^{3+}
Gaya I (250 km SE)*							
Plateau	AhBsc	6.0	1.8	0.4	<0.1	8.2	7.6
Foot of slope	A1	5.7	1.9	0.5	0.8	9.1	2.8
	A2	3.8	1.0	0.3	0.3	5.4	2.5
	2Bs1	<0.1	<0.1	0.1	0.1	0.2	1.7
	2Bs2	<0.1	<0.1	0.2	0.3	0.5	1.9
Depression	A1	2.4	0.8	0.5	<0.1	3.7	1.7
	A2	2.4	0.9	0.3	<0.1	3.6	1.4
	2Bs1	0.4	0.2	0.2	<0.1	0.8	1.5
	2Bs2	0.9	0.1	0.1	<0.1	1.1	1.1
Kobadié (60 km SSW)							
Plateau	A	17.3	9.9	<0.1	<0.1	27.2	4.3
	2Bsc	10.3	3.7	<0.1	<0.1	14.0	17.6
	2Bsc2	5.9	2.5	<0.1	<0.1	8.4	41.7
Foot of slope	crust	0.9	0.3	0.5	0.1	1.8	3.0
	A	1.4	1.0	0.5	0.1	3.0	2.9
	C	2.1	0.9	0.5	0.2	3.7	1.8
	C2Bs	1.4	1.5	0.3	0.3	3.5	1.3
Depression	A1	22.4	2.5	6.0	0.2	31.1	<0.1
	A2	11.5	2.1	2.9	0.2	16.7	<0.1
	Bw1	18.0	2.1	2.9	0.2	23.2	<0.1
	Bw2	18.4	1.9	1.8	0.5	20.1	<0.1
Gagani near Sadoré (35 km SSE)							
Plateau	A	2.0	0.5	0.6	<0.1	3.1	3.5
	2Bsc1	2.9	1.0	<0.1	<0.1	3.9	3.0
	2Bsc2	5.3	1.8	<0.1	<0.1	7.1	2.7

(*Source*: Hammer 1994, summarized)
*Distance and compass direction from Niamey

Table 1.2–5. Quantities of organic matter (in kg m^{-2}) in a dune catena in southwestern Niger at four locations. Values are calculated for 1 m profile depth

Catena	Sadoré	Konébéri	Ouallam	Ayorou
Plateau	3.4	1.2	2.9	1.7
Depression	4.9	4.4	3.3	3.4
Sum*	8.3	6.6	6.2	5.1

(*Source*: Hammer 1994, summarized)
*Theoretical value: sample sites (plateau, depression) were summed up for better comparison

Abstract

Sandy soils are widespread in Niger and originate primarily from the two dune systems of the Upper Pleistocene epoch. The older system (more than 40 000 years old) forms the soils of south-western Niger. Clay translocation in the soil profile (Luvic Arenosols) has increased the sensitivity of the topsoil to erosion. The younger dune system (16 000 to 20 000 years old) is located at the southern end along the banks of the Niger River. These soils are weathered to a small or moderate extent (Dystric Regosols and Cambic Arenosols, respectively) and may receive crusted material from laterite steps (Chromic Cambisols). In the lower slope positions of inselbergs and steps, mixed dune deposits were formed (Ferralic Arenosols) being more sensitive to erosion by water than by wind. While Ferralsols dominate the upper slope positions, Leptosols are found on the plateaus. Additionally, shallow unweathered sandy soils (Eutric Regosols) and weakly weathered outcrops from metamorphic rocks (Calcaric Regosols) occur in western Niger. In the depressions, Eutric and Dystric Gleysols as well as Orthic Solonchaks (salt depressions) developed.

Wind-blown soil at the beginning of the rainy season can affect the soil surface by grinding it. In river valleys and gullies, sand deposition can also take place. During the dry season, strong 'Harmattan' winds transport silty dust from eastern Niger into western Niger. Deposited silt contributes to surface crusts in flat depressions, thus, decreasing infiltration as well as increasing runoff and water erosion on these sites.

Résumé

Les sols de l'ouest du Niger

Les sols sableux sont très répandus au Niger et essentiellement hérités des deux systèmes dunaires du Pleistocène supérieur. Le système le plus ancien (datant d'il y a plus de 40 000 années) constitue la surface du sud-ouest du Niger. La pénétration du sous-sol par l'argile (Arénosols luviques) a augmenté la sensibilité de la couche supérieure du sol à l'érosion. Le système dunaire le plus récent (16 000 à 20 000 années) longe les berges de la rivière Niger dans la partie sud de sa distribution. Ces sols sont plus ou moins altérés (respectivement Régosols dystriques et Arénosols chromiques) et reçoivent éventuellement des particules de ressauts de latérite (Cambisols chromiques). Sur les pentes inférieures des affleurements et des ressauts, des dépôts mixtes de dune (Arénosols ferraliques) se sont formés, plus sensibles à l'érosion par l'eau qu'à celle du vent. Pendant que les Ferralsols prédominent sur les pentes supérieures, on trouve des Leptosols lithiques sur les plateaux. On trouve également des sols sableux superficiels non altérés (Régosols eutriques) et des affleurements faiblement altérés de rochers métamorphiques (Régosols calcariques) à l'ouest du Niger. Dans les dépressions, des Gleysols eutriques et dystriques ainsi que des Solonchaks orthiques (dépressions de sel) se sont formés.

Au début de la saison des pluies, le sable transporté par le vent peut affecter la surface du sol en la pulvérisant. Dans les vallées des rivières et les ravins, il arrive aussi que le sable se dépose. Pendant la saison sèche les forts vents d''harmattan' transportent de la poussière de limon de l'est à l'ouest du Niger. La boue déposée contribue à la formation de croûtes à la surface des dépressions plates, limitant ainsi l'infiltration tout en favorisant l'écoulement et l'érosion par l'eau sur ces sites.

References

Bationo, A., S.H. Chien, J. Henao, B.C. Christianson, and A.U. Mokwunye. 1990. Agronomic evaluation of two unacidulated and partially acidulated phosphate rocks indigenous to Niger. *Soil Science Society America Journal* 54: 1772–1777.

Chen, Y., J. Tarchitzky, J. Brouwer, J. Morin, and A. Banin. 1980. Scanning electron microscope observations on soil crusts and their formation. *Soil Science* 130: 45–55.

FAO. 1988. Soil map of the world. FAO-UNESCO, Paris, France.

Greigert, J. 1966. Description des formations cretacées et tertiaires du Bassin des Jullemmeden (Afrique occidentale). Editions du Bureau de Recherches Géologiques et Minières, Paris, France.

Hammer, R. 1994. Bodensequenzen und Standortseigenschaften im Südwest-Niger/West-afrika. Hohenheimer Bodenkundliche Hefte 22, Stuttgart, Germany.

Hoogmoed, W.B. and L. Stroosnijder. 1984. Crust formation on sandy soils in the Sahel. I. Rainfall and infiltration. *Soil and Tillage Research* 4: 5–23.

Völkel, J. 1989. Geomorphologische und pedologische Untersuchungen zum jungquartären Klimawandel in den Dünengebieten Ost-Nigers (Südsahara und Sahel). Bonner Geographische Abhandlungen 79, Bonn, Germany.

1.3
Farming Systems in Niger

Jojo Baidu-Forson and Timothy O. Williams

Introduction

Farming systems in Niger can generally be characterized as incorporating combinations of cereals, legumes, livestock, and trees. The basic components are a pastoral livestock husbandry and an integrated sedentary crop-livestock system. Naturally growing trees are widely used for firewood, fodder, and construction material (Chapter 4.3). However in some areas, tree stocks have greatly decreased due to heavy pruning (Rinaudo 1996). Generally, Nigerien farmers do not plant trees except occasional fruit trees; but they tolerate certain tree species for special purposes, for example wind erosion control (Chapter 4.3).

Consistent with what has been observed elsewhere in the West African semi-arid Tropics (Norman *et al.* 1981), the primary objective of the predominantly subsistence and semi-subsistence farm households in Niger is the assurance of food supplies. Strategies used by Nigerien farm households to achieve this primary objective include: diversification of crop enterprises including, where possible, the use of different tree species (Plate 8); off-farm income generation activities including commerce and sale of firewood; off-season migration to coastal countries; and sale of livestock. Sales of crop residues and herbaceous plants collected from communal rangeland also occur in some regions (Baidu-Forson 1994). These strategies permit households to minimize the risk of failing to meet subsistence food requirements by diversifying sources of income and food.

Major Crops

The predominant cropping system in Niger is millet-based (Plate 9). Sole millet and millet-based intercrops cover at least 90% of the total cultivated area. Most of the cereals are reserved for subsistence consumption, although

B. Buerkert, B. E. Allison and M. von Oppen (eds.), Wind Erosion in Niger, 33–42.

periodic sales occur when money is needed to meet family obligations. The main cash crops are cowpea, groundnut, onions, and cotton. Generally, the number and types of crops grown are limited by crop growth potentials of the different agroecological zones of Niger, which are determined by climatic conditions, soil, and water resources. The number of crops increase southward from the arid northern Sahel to the more humid northern Sudanian zone in the Gaya district. Crop diversification is lowest in the northern Sahel because the harsh environment and short growing season limit the types of crops that can successfully be grown. However, the longer growing season and wetter conditions in the southern Sahelian to northern Sudanian zones offer greater opportunities than other regions in Niger for the cultivation of many more staple (millet, sorghum, some small amounts of maize) and cash crops (primarily cotton and groundnut).

Analysis of Production of Principal Crops

The principal crops in Niger are millet, cowpea, and groundnut. Changes in average levels of crop production, area cultivated, and per capita production of these crops are analyzed for the periods 1982–1984 and 1989–1991. As the percentage increments in millet production were less than the percentage increases in area cultivated, both nationally and in the Tillabéry region (western Niger), area expansion is likely to be the main source of increased overall production (Table 1.3–1). However, in the Diffa region (eastern Niger) the increment in area cultivated with millet was less than the increase in production, possibly due to better yield gains. The increases in production, both nationally and in the Diffa and Tillabéry regions, may be partly due to the relatively better rainfall in the 1989–1991 period compared with the 1982–1984 period during which drought occurred in Niger. The increase in millet production in the Diffa region (141%) was particularly larger than the increase recorded in the Tillabéry region (6%), albeit from a lower production base. Nationally, a slight decline occurred in the per capita production. This suggests that the increase in millet production did not match the increase in population. Regionally, large increases in per capita production occurred in the Diffa region, while a 17% decrease occurred in the Tillabéry region.

Cowpea area, production, and per capita production increased both nationally and in the two extreme regions of Niger. The largest increase in per capita production occurred in the Diffa region, possibly as a result of proximity to important markets for grains (particularly cowpea) in nearby Nigeria. However, considerable declines occurred in area cultivated, production, and per capita production of groundnut in Niger (Table 1.3–1). This general decline over the periods analyzed reflects declining interest in groundnut production, which is attributed to: the cessation of groundnut purchase operations by a government agency, the declining length of the growing season (Chapter 1.1)

Table 1.3–1. Changes in cultivated area, production and per capita production of major crops from 1982–1984 to 1989–1991 in Niger

Crop	Niger		Eastern Niger Diffa region		Western Niger Tillabéry region	
	1982–1984 Average	1989–1991 % change*	1982–1984 Average	1989–1991 % change	1982–1984 Average	1989–1991 % change
Millet						
Area (10^3 ha)	3081.7	+36	38.4	+128	793.0	+10
Production (10^3 mt)	1124.9	+27	9.0	+141	279.7	+6
Per capita production (kg per capita)	184	-0.5	57	+87.7	192	-16.7
Cowpea						
Area (10^3 ha)	1477.6	+67.6	22.1	+8.5	261.3	+28
Production (10^3 mt)	227.2	+43	3.7	+185	23.3	+68
Per capita production (kg per capita)	37	+13.5	23	+126	16	+31.3
Groundnut						
Area (10^3 ha)	166.7	-55.3	n.a.	n.a.	3.9	+12
Production (10^3 mt)	64.4	-58.0	n.a.	n.a.	1.6	+22
Per capita production (kg per capita)	11	-72.5	n.a.	n.a.	1.0	0

(*Source*: Ministère de l'Agriculture et de l'Elevage 1993)
*This represents the percentage increase or decline of corresponding data for 1989–1991 average when compared with the 1982–1984 average
n.a. = not available

and incidence of rosette disease. Although the Tillabéry region recorded increases in both groundnut production and area cultivated with groundnut, it is not a major groundnut producing region.

Major Biophysical Constraints, Fertility, and Crop Management

A constraint to crop production in the southern Sahelian zone is the low soil fertility (Bationo and Mokwunye 1991). Additionally, strong convective storms annually cause wind erosion of topsoils. The eroded topsoils eventually cause damage to young millet plants and losses in soil fertility. Traditionally, farmers in Niger relied on long fallow periods for regeneration of the fertility of the poor soils. However, fallow practices, particularly on lands under communal ownership regimes, have changed in response to increased pressure on arable land. There is a general trend to reduce the length of fallow periods.

The lengths of time for which land can be allowed to remain in fallow depend on the intensity of land pressure, land ownership status and productivity of the land. Surveys conducted in the Hamdallaye district of western Niger, show that fallow periods last for two to three years on borrowed land and three to five years on owned land, instead of the former general practice of ten years or more (Taylor-Powell *et al.* 1991). The same study shows that 35% of sample farmers leave portions of their fields in fallow instead of putting whole fields under fallow. This is a mechanism employed to ensure land-use rights are not lost.

Organic matter from livestock manure, compound wastes, and crop residues are the major sources of soil amendments in Niger. Areas around household compounds receive most manure and urine from penned livestock and household refuse. This pattern of organic matter deposition creates a fertility gradient determined by the distance of plot location from farm household compounds (Plate 10), similar to what was found in Burkina Faso (Prudencio 1987, 1993). Depending on crop value and the responsiveness to improved fertility, farm households allocate different crops to locations on the fertility gradient. If soil and moisture conditions permit, higher-valued crops such as maize, sorghum, and vegetables are allocated to fields where most of the manure and household refuse are dumped. Where households have access to transportation, some manure and household refuse are carted to millet fields, particularly in the densely populated districts around Maradi and Birni N'Konni.

Farmers are aware of the steady decline in soil fertility and try to counteract it by applying cut bushes and millet crop residues on selected poor or crusted spots of their fields (Chapter 4.3). Crop residues represent an important option for recycling nutrients, replenishing organic matter, and protecting soils from effects of wind erosion (Chapters 3.3 and 3.4). Yet, research conducted in Niger showed that at the beginning of a new cropping season, the amount of straw remaining on farmers' fields averaged only: 100 kg ha^{-1} in the Hamdallaye district (east of Niamey), 150 kg ha^{-1} in the Dantiandou district (southeast of Hamdallaye), and about 250 kg ha^{-1} in the Kirtachi district, located in the transition area between the Sahelian and northern Sudanian zones (Baidu-Forson 1995). The low amount of crop residues is linked to the low production of straw, particularly in the drier northern districts. Also, the amount of straw that remains on fields declines with distance from the household compound, possibly as a result of a fertility gradient (Baidu-Forson 1995).

The use of inorganic fertilizer amendments, especially phosphorus, to increase crop production is minimal. According to the Food and Agricultural Organization of the United Nations (FAO 1993), the annual application of mineral fertilizers in Niger is 0.3 kg ha^{-1}, one of the lowest in the world. Since supply is a major constraint in the use of organic soil amendments, the application of inorganic fertilizers represents an important option for increasing biomass production. This is feasible under certain conditions (Chapter 4.2).

Variability across Toposequence

Studies indicate that physical and chemical soil properties can vary with toposequence location (Van Staveren and Stoop 1985; Chapter 1.2). However, landscape position is not a major factor in land-use by local farmers at Hamdallaye (east of Niamey) (Ouattara 1990). The main decisive factors in land-use at Hamdallaye are: surface texture, surface degradation, crusting, and decline in land productivity. Surveys conducted at Malgorou (north of Gaya) showed that farmers plant different crops depending on toposequence position as well as the soil type (Ouattara 1990). The differences in land allocation patterns may be partly explained by differences in soil characteristics as well as diversity of crops that can be grown.

Plateau areas of a typical toposequence in Niger have topsoils consisting of gravel and laterite (Chapter 1.2) and are frequently left to natural vegetation and livestock grazing. Where plateau areas are cultivated, soils are shallow and sandy with numerous lateritic outcrops (McIntire *et al.* 1989). Generally, soils on mid-to-lower slopes have deep sandy soil profiles (Chapter 1.2) and are cultivated with millet and millet-based intercrops while in the extreme lower portions of the toposequence, particularly along the valley of the Niger River where Gleysols and Vertisols dominate (Chapter 1.2), some sorghum and rice are cultivated.

Farm Size and Labor

Farm sizes generally decline southward, primarily as a function of population density. Average farm size in the thinly populated northern Sahelian region of Zarmaganda (east of Tillabéry) was estimated (1982–1985) at 10 to 15 ha per household, approximately 1.3 to 1.8 ha per adult equivalent (McIntire *et al.* 1989). By contrast, farm sizes average between 1 and 1.5 ha per adult equivalent in the relatively more densely populated southern Sahelian region of Boboye. Average farm sizes are much smaller in the even more densely populated areas in the Maradi region. For example, median farm size for a family of seven is only about 4.6 ha in the densely populated Madarounfa district (Swinton 1985).

Family labor is the predominant source of farm labor while collective labor (*borgu*) or invitational labor (*gaya*) provide assistance to overcome peak labor demands at weeding and harvesting. Studies conducted in Niger have estimated the total labor hours used per ha at between 152 and 245 (Legal 1985; Swinton 1988a; McIntire *et al.* 1989). Spatial and interannual variability in labor demands occur, primarily as a result of differences in soil type and rainfall. The distribution of total time spent on farming activities has been estimated as: 6 to 11% for field preparations, 8 to 14% for planting, 59 to 69% for weeding, and 14 to 29% for harvesting (McIntire *et al.* 1989).

The Role of Livestock in the Farming Systems

Livestock play very important roles in the farming systems of Niger. Pastoralism, i.e. livestock rearing without crop production, has always been a way of life for most of the inhabitants of the northern part of the country (annual rainfall $\leq$ 300 mm; Chapter 1.1). Here, crop cultivation cannot be undertaken, but seasonal pastures exist which livestock can exploit. In this region, pastoral production represents a rational and profitable system of land-use. With more rainfall further south, mixed farming in the sense of keeping goats, sheep, and cattle on crop farms is common. Many farmers accumulate and use livestock as productive assets to reduce risk and mitigate the effects of drier than average years. Swinton (1988b) found that livestock are the most prevalent, most liquid and potentially the most productive assets of agricultural households in Niger. Livestock serve as a credit-line or cash generator for seasonal requirements of agricultural activity. After a good crop harvest, animals provide an important investment opportunity for surplus funds, while in poor crop production years they are sold to buy food for household sustenance.

Livestock are also kept to complement cropping activities through the provision of manure and traction power. Animal manure is an integral component of soil fertility management practices in Niger (Chapter 4.3). The low level of inorganic fertilizer use in the country implies that as population pressure increases and fallow cycles are shortened, animal manure becomes one of the principal sources of nutrients for soil fertility maintenance and crop production.

Table 1.3–2 shows the mean and range of household livestock holdings in two villages in western Niger in 1983 and 1991. It indicates the predominance of small ruminants over cattle even before the last major drought of 1984–1985. Since the drought, small ruminants have become even more important. On a national level, the relative decline in cattle compared to sheep and goats was much smaller, although the absolute numbers of cattle decreased drastically. The number of donkeys – the main species used for transportation of farm produce – declined from about 492 000 in 1983 to 448 000 in 1991 (Ministère de l'Agriculture et de l'Elevage 1993).

In sedentary farming communities, women own most of the small ruminants while men own most of the cattle. Decisions on the sale and purchase of animals rest with individual adult owners, but the entire household herd is often herded together and managed as a single entity for grazing and land-manuring purposes.

Table 1.3–2. Mean livestock holdings per farm household in two villages in the Ouallam region (western Niger) in 1983 and 1991

	Sadeize-Koara (n = 15)				Samari (n = 25)			
	1983		1991		1983		1991	
Animal type	Mean	Range	Mean	Range	Mean	Range	Mean	Range
Cattle	5.5	0–27	1.8	0–10	2.3	0–14	1.0	0–5
Sheep	6.6	0–29	9.6	0–32	1.4	0–8	6.4	0–17
Goats	12.5	0–40	22.3	1–81	10.6	0–30	16.3	2–60

(*Source*: Data 1983, M. Dicko (unpublished data); data 1991, ILCA 1993)

In Niger, the adoption of animal traction for farm operations has been limited. Although its use has been promoted since the 1950s, sustained adoption has occurred in relatively limited locations such as low lying valleys and irrigated rice schemes. Elsewhere in the country not only have initial adoption rates been low, but early adopters have often stopped using animal traction after three to five years. The low demand for animal draft power in some parts of Niger can be attributed to the absence of agronomic, demographic, and market conditions under which animal traction has thrived in other parts of sub-Saharan Africa (Pingali *et al.* 1987). Such conditions include (i) a high proportion of heavy soils which are difficult to work manually, (ii) a long preparatory rainfall period to reduce the labor conflict between plowing and timely planting, (iii) crops with high yield response to improved soil preparation, and (iv) market conditions which ensure an elastic demand for increased production. In other areas, the constraints which have limited sustained adoption of animal traction include inappropriately designed equipment, inadequate support services, lack of credit, low farm gate prices, and high equipment costs. Together these constraints reduce the profitability of animal traction (Tahirou *et al.* 1993; Williams 1994). Nonetheless, in areas where draft animals are used for plowing and weeding, oxen are generally preferred, but where animals are used only for weeding, donkeys are most likely to be employed.

In addition to tillage, most farmers also use their draft animals to transport manure on carts to outlying fields and crop residues to markets. The hiring-out of animals and carts is common and this represents a significant source of income for farmers who own animal traction equipment, particularly in dry areas. A survey conducted in Samari, Ouallam region, showed that transport services accounted for 70 to 80% of the income derived from such rental activities (Williams 1994).

Since the early 1980s, in most parts of the Sahel including Niger, many transhumant herds have moved southward to areas where natural pastures and crop residues are more abundant. This increasing concentration of stock in cropped areas has, in some cases, depleted the vegetative cover (either through grazing and/or harvesting of crop residues for feed) and left soil surfaces bare and more susceptible to wind erosion. The extension of arable farming to marginal lands unsuitable for cultivation has also deprived transhumant herds of traditional grazing lands and has exacerbated the problem of land degradation. The challenge in these areas is to develop ecologically sound and economically viable management practices for crop and livestock production.

Migration

In times of food deficits and during the long dry season, there is considerable migration out of Niger to coastal countries such as Nigeria, Ghana, and Ivory Coast. The migration may be temporary (Swinton 1988b), to seek off-season wage labor, or permanent in response to chronic food deficits. Remittances from migrants contribute to the assurance of food security for non-migrating family members. Although migration has historically been limited to young men, recent migrations include women.

The extent to which migratory income is directly invested in agriculture, in general, and erosion control, in particular, is not clear. Generally, migrations take place between October and January, particularly right after the end of the cropping season. Since the post-harvest period is the only time when labor would be available to implement erosion control techniques, seasonal migrations at this time contribute to labor shortages. With the migration of able-bodied young men, usually only women and a few old men are left in the villages. This directly limits the type and scale of erosion control techniques that can be implemented.

Conclusions

With increasing population pressure, the low input and extensive farming systems of Niger are slowly becoming intensive and sedentary. Millet-based crop and livestock production systems are likely to predominate. The per capita production of the major food and cash crops need to be raised considerably. This will require the exploitation of opportunities for complementary integration of crops and livestock, improvement of soil fertility status from both inorganic and organic sources, and control of biophysical constraints such as wind erosion. Labor constraints at weeding and for implementation of erosion control techniques, occur due to dependence on family labor and seasonal migration. They may be resolved with simple and cheap labor-saving technologies.

Abstract

Farming systems in Niger can generally be characterized as incorporating combinations of cereals, legumes, livestock, and trees. Low soil fertility and wind erosion affect crop productivity. Recently, traditionally long fallow periods have been replaced by much shorter fallow periods. The shorter fallow periods could be compensated for by fertility improvement through organic matter application. However, organic matter is often limited in supply. Average levels of cultivated area and production for millet, cowpea, and groundnut between the periods 1982–1984 and 1989–1991 have changed. Farm sizes generally decline southward and family labor is the predominant source of farm labor. Livestock plays a very important role in the farming systems. Adoption of animal traction for farm operations in Niger is limited, principally due to the absence of agronomic, demographic, and market conditions that would favor widespread use of animal traction.

Résumé

Systèmes d'exploitation agricoles au Niger

Les systèmes d'exploitation agricoles au Niger peuvent généralement être caractérisés comme des systèmes intégrant céréales, légumineuses, élevage et arbres. La faible fertilité du sol et l'érosion éolienne affectent le rendement des récoltes. Récemment, les longues périodes de jachères traditionnelles ont été remplacées par des jachères beaucoup plus courtes. On peut compenser ces jachères courtes en améliorant la fertilité des sols par l'application de matière organique. On est souvent limité cependant quant à son approvisionnement. Les moyennes d'espaces cultivés et les productions de petit mil, de niébé et d'arachide entre les périodes 1982–1984 et 1989–1991 ont changé. La taille des exploitations diminue généralement lorsque l'on se déplace vers le sud, et la main d'oeuvre familiale est la principale source de force du travail dans les activités agricoles. L'utilisation de la traction animale pour les travaux agricoles au Niger est limitée, principalement en raison de l'absence des conditions agronomiques, démographiques et du marché qui favoriseraient son utilisation à grande échelle.

References

Baidu-Forson, J. 1994. Uses and economic value of crop residues and herbaceous plants to rural households in western Niger. Resource Management Program, Progress Report No. 5, ICRISAT Sahelian Center, Niamey, Niger.

Baidu-Forson, J. 1995. Determinants of the availability of adequate millet stover for sustainable crop production in the Sahel. *Journal of Sustainable Agriculture* 5(1/2): 101–116.

Bationo, A. and A.U. Mokwunye. 1991. The role of manure and crop residues in alleviating soil fertility constraints to crop production: With special reference to the Sahelian and Sudanian zones of West Africa. *Fertilizer Research* 29: 117–125.

FAO. 1993. AGROSTAT-PC. Computerized Information Series. Statistics Land Use 1/3, FAO, Rome, Italy.

Legal, P.Y. 1985. La gestion de la force de travail sur 14 exploitations agricoles Haoussas. Programme de Recherche-Développement, Projet de Développement Rural de Maradi, Niger.

McIntire, J., J. Hopkins, J. Lamers, and L.K. Fussell. 1989. The millet systems of western Niger. I. Crop production. ILCA, Addis Ababa, Ethiopia.

Ministère de l'Agriculture et de l'Elevage. 1993. Annuaire des statistiques de l'agriculture et de l'élevage 1991. Service d'Analyse des Politiques et de la Coordination des Statistiques, Direction des Etudes et de la Programmation, Niamey, Niger.

Norman, D.W., M. Newman, and I. Ouedraogo. 1981. Farm and village production systems in the semi-arid Tropics of West Africa: An interpretive review of research. Research Bulletin No. 4, ICRISAT, Hyderabad, India.

Ouattara, M. 1990. A study of two toposequences of the dry valley systems of western Niger (West Africa). Ph.D. Diss. Texas A&M University, TX (unpublished).

Pingali, P.L., Y. Bigot, and H.P. Binswanger. 1987. Agricultural mechanization and the evolution of farming systems in sub-Saharan Africa. Johns Hopkins University Press, Baltimore, MD.

Prudencio, Y.C. 1987. Soil and crop management in selected farming systems of Burkina Faso. OAU/SAFGRAD, Ouagadougou, Burkina Faso. 152 p.

Prudencio, Y.C. 1993. Ring management of soils and crops in the West African semi-arid Tropics: The case of the Mossi farming system in Burkina Faso. *Agriculture, Ecosystem and Environment* 47: 237–264.

Rinaudo, T. 1996. Tailoring wind erosion control measures to farmers specific needs. pp. 161–171. *In* B. Buerkert, B.E. Allison, and M. von Oppen (eds.) Wind erosion in West Africa: The problem and its control. Proceedings International Symposium, Hohenheim, Germany, 5–7 Dec. 1994, Margraf Verlag, Weikersheim, Germany.

Swinton, S.M. 1985. Les ressources agricole des exploitations dans trois villages représentatives de l'arrondisement de Madarounfa en 1984. Programme de recherche sur les systèmes de production. Document No. 7, Institut National de Recherche Agronomiques du Niger (INRAN), Niamey, Niger.

Swinton, S.M. 1988a. Cropping systems budgets for south-central Niger in 1984 and 1985. University of Minnesota, Department of Agricultural Economics, St. Paul, MN.

Swinton, S.M. 1988b. Drought survival tactics of subsistence farmers in Niger. *Human Ecology* 16: 123–144.

Tahirou, A., J. Lowenberg-DeBoer, and F. Kruit. 1993. Enquête sur la rentabilité de la culture attelée dans une zone sahélienne. Document No. 25F, Département de Recherches en Economie Rural, Institut National de Recherches Agronomiques du Niger (INRAN), Niamey, Niger.

Taylor-Powell, E., A. Manu, S.C. Geiger, M. Ouattara, and A.S.R. Juo. 1991. Integrated management of agricultural watersheds: Land tenure and indigenous knowledge of soil and crop management. TropSoils Bulletin No. 91–04, Soil Management Collaborative Research Support Program, North Carolina State University, Raleigh, NC.

Van Staveren, J.P. and W.A. Stoop. 1985. Adaptation of sorghum, millet and maize to the different land types in a common toposequence in West Africa: Implications for the testing and introduction of new sorghum varieties in the local agriculture. *Field Crops Research* 11: 13–35.

Williams, T.O. 1994. Problems and prospects in the utilization of animal traction in Niger. ILCA Semi-Arid Zone Programme, Niger (mimeo).

PART TWO

Extent and Implications of Wind Erosion and Dust Transport

2.1 Origin, Deposition, and Composition of Dust, and Consequences for Soil and Site Properties with Special Reference to the Semi-arid Regions of West Africa

Karl Stahr and Ludger Herrmann

Problem Definition

The Sahara desert is the most important source area for soil-derived dust in the world. Between 60 and 200 million tons of dust per year are generated from this area and contribute to the dust load of the troposphere (Morales 1979). In general, dust generation is a natural process which may be modified in the desert fringe areas by human activities such as livestock grazing, wood cutting, and soil tillage practices. The influence of dust on soils is well-known in areas of glacial loess, such as the Ukraine or the Great Plains in Canada, as well as in areas characterized by desert loess like China or Israel. In the last decade, it became obvious that dust was also transported to other areas in reasonable quantities where it formed admixtures to the soil, modified local weather conditions, directly affected vegetation, or had long-term effects on soil properties (Morales 1979; Pye 1987; Jahn 1995). When studying dust transport, it is not only necessary to consider the loss from one area and the sedimentation in another area, but also the redistribution of dust particles. Therefore, a total balance of dust fluxes for a particular site can only be assessed when all accreting and generating events can be calculated.

All dusts have certain properties of grain-size distribution, mineralogy, and chemistry. These are generated from the dust sources far away but are also affected by local admixtures, as shown with the loess of south-west Germany (Alaily 1972; Maus and Stahr 1977). A permanent and long lasting influence of dust on the properties of ecosystems is expected in areas where the local parent rocks and soils have properties differing in one or more

B. Buerkert, B. E. Allison and M. von Oppen (eds.), Wind Erosion in Niger, 45–65.

parameters from those of the imported dust. As far as grain size, mineralogy, and chemistry are concerned, dust may influence the water, nutrient, and air regime of a given soil. This is especially the case in Sahelian zone countries like Niger where relatively coarse grained and mineral-poor soils (Arenosols) derived from the so-called *Continental Terminal* as well as from local dune sands (Chapter 1.2) prevail. These soils have a low water holding capacity and insufficient potassium (K) and phosphorus (P) supplies (Hammer 1994) and could well be improved by dust addition. In view of this hypothesis, Wilke *et al.* (1984) and Drees *et al.* (1993) tried to quantify the effect of dust addition in soils from Nigeria and Niger. This paper aims to give a review of the current knowledge on quantification of the effect of dust deposition on Sahelian soils. Special emphasis is given to the study area Niger.

Dust is defined as soil-derived solid particles with a diameter of less than 63 μm, which have been or are currently transported in suspension in the air. As far as soils are concerned, the fraction of less than 63 μm is taken as the 'potential dust fraction'.

Dust Transport Process

The dust transport process is a typical translocation of soil masses with a generally horizontal vector. Like all translocation processes it can be subdivided into three subprocesses. These are (i) the generation of dust particles as well as their release into the air, followed by (ii) the real transport process in suspension, and finally (iii) the sedimentation or accretion of this dust. After sedimentation, dust particles may be remobilized or incorporated into the soil matrix, thereby influencing the permanent properties of the soil in question.

Erosion – Dust Generation

A prerequisite for dust transport is the generation of dust-size particles. Silt-size particles are generated mainly by physical weathering. In arid areas, this is temperature cracking or salt weathering. Furthermore, physical impact on large crystals can lead to crystalline breakage during saltation of sand-size particles. In addition to these physical processes, chemical weathering may also result in a diminution of larger crystals into the silt size. Some of the silt-size particles may be of biological origin such as opal phytoliths and calcareous and siliceous remnants of algae. Others may be composites. The latter occurs if clay aggregates and pellets break down into silt size, or clay coatings cover small crystals of feldspars, micas, quartz, and carbonates. All these processes which generate silt-size particles are of importance in the central and south Saharan environment (Pye 1987; Gomes *et al.* 1990). Besides these processes relevant in tropical deserts and tropical desert fringe areas, glacial grinding, frost weathering, and fluvial comminution also occur as processes that generate dust particles.

Taking into account that dust storms occur several times a year, the generation of dust particles from soil surfaces, even in deserts, may not be rapid enough to produce sufficient erodible particles to load these dust storms. Therefore, as a regular prerequisite, an enrichment of silt-size particles should take place prior to erosion during a dust storm event. This concentration of dust-size particles is often a result of limited water flows into hollows and along stream lines. Of minor importance for the area in question are glacial outwash areas which may be major sources for the periglacial loess. Since loess landscapes are mainly found in the subtropical and temperate areas, they cannot be source areas for the sub-Saharan tropical dust. Equally of minor importance for the dust uptake into the air are stony deserts (*Hamadas*), gravely deserts (*Serirs*), and *Erg* dune fields. On the other hand, argillaceous and silty bedrocks, shales, and slates as well as deeply weathered saprolites and old palaeosol surfaces may generate relatively high amounts of dust particles by physical weathering.

Since millions of tons of sediments must be ready for entrainment, only those areas which are regularly covered by fine new sediments can be counted as major source areas, especially for the Sahel. According to Yaalon and Ganor (1973), wadi sediments, or in other words alluvial flood-plains of deserts, are the major source areas. Coastal sabkha sediments could be major sources, although they are not widespread in the area in question. Fossil and recent lake sediments of Lake Chad, as a terminal lake, can also serve as a primary or secondary source (Felix-Henningsen 1992). Finally, alluvial fan, lake, and playa sediments (Mainguet 1991), derived from the major inner Saharan mountain areas of the Tibesti, the Hoggar, and the Aïr, play an important role (Herrmann *et al.* 1994b). These alluvial fans surround the mountainous regions, mainly southward. The importance of intermediate sources, where dust is remobilized from a dune field, are still in question (Völkel 1988).

The surfaces must be suitable for the aeolian uptake, i.e. silt-size particles should not be aggregated and clay-size particles not formed into aggregates too large to be suspended. Average storms only receive dust loads of relevance if the surfaces are dry, even, and not covered with vegetation or stone pavements. All these prerequisites are fulfilled in the superarid areas surrounding the Saharan mountains (Pye 1987; Bergametti 1992). The more favorable the erodibility of soil surfaces the easier is the uptake of silt-size particles by the wind, and the less is the necessary erosivity of winds and storms (Wilson and Cooke 1980). In spite of the fact that some easily available sources do exist in the southern Saharan zone, annually only 10 to 20 sandstorms are observed. That means that in the zone, favorable conditions for uptake and transport prevail more often than dust events really occur.

Transport – Aerosols and Wind Systems

Wind transported material may move in three different modes: creep, saltation, and suspension (Pye 1987). The transportation mode varies with regard to wind speed and properties of the particles. The differences in wind speed are less pronounced than the differences in grain size, grain shape, and particle density. The latter difference mainly governs the transport process, as dust particles may first be separated from the soil surface by creep and saltation, often in combination with aggregates or coarser particles. However, during the transport process they may be liberated from their host particles and then they can be more easily lifted to higher altitudes by vertical air movement. Depending on the settling velocity and upward wind movement, particles smaller than 100 and 50 μm in diameter can be transported in suspension over short or long distances, respectively. As soon as wind speed is reduced, the diameter and quantity of the material transported decreases (Pye 1987; Figs. 2.1–1 and 2.1–2).

Dust is transported over long distances as soon as it reaches heights which are no longer affected by the roughness of the ground (Morales 1979). Dust is often lifted by warm uplift currents. Frequently, where steep pressure gradients exist in the atmosphere, cold down draught haboobs may lift particles into the air. As soon as dust is involved in convective updraughts like 'dust devils' and thunderstorm cells, it may be carried to great heights. Then it can be transported over distances of thousands of kilometers by jet streams within a rather short period. As observed from work in the United States, dust storms are often followed by rain events within a few hours (Pye 1987). This observation can also be made in the Sahel. There, at the beginning of the rainy season, convective storms travel along the intertropical convergence zone (ITCZ). During these storm events, suspended dust particles are absorbed by rain droplets and then transported back to the earth's surface.

During transport, a change of chemical and mineralogical properties has been observed which may be induced by gaseous, liquid, or solid state chemical reactions in the aerosols (Singer 1986). One of the reactions reported by Bergametti (1992) is the dissolution of calcium carbonate in the presence of sulfur oxide/sulfuric acid. This leads to the formation of gypsum particles. Another change, important for the properties of the incoming dust for the soil, is the uptake of K from sea sprays into smectitic or mixed layer clay minerals. This leads to advanced illitisation of the dust-transported clay particles (Singer 1986). These reactions are more likely when condensed water particles take up gaseous molecules or ionized particles from the air.

Deposition – Fallout, Rainout, Accretion

As far as the influence of dust on soils is concerned, only permanently deposited dust loads are of interest. According to current knowledge, four ways of

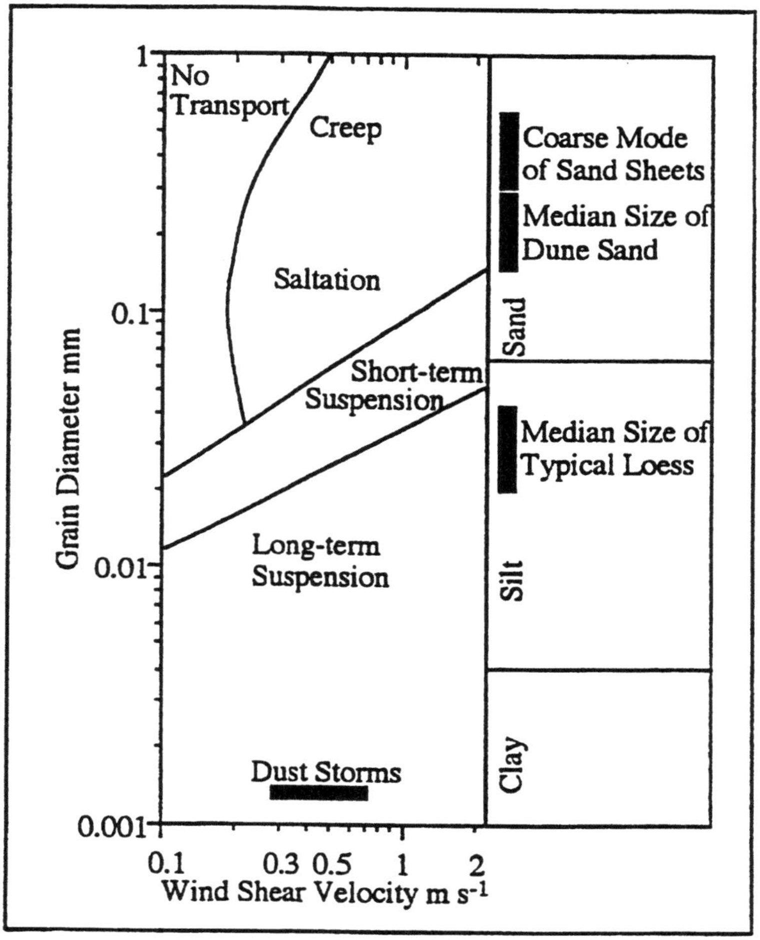

Fig. 2.1–1. Relation between transfer mode, grain diameter, and wind speed. (*Source*: Pye 1987, modified).

dust deposition have been described (Pye 1987). The simplest deposition mode is fallout from the atmosphere because of decreasing wind velocity and turbulence. The boundary condition is that the upward directed wind vector must be less than the sinking velocity of dust particles in order to enable them to reach the soil surface. For example, a particle with a diameter of 50 μm will sink when the wind velocity is below 2 m s^{-1}, while a particle of 10 μm only reaches the ground when the velocity is below 0.1 m s^{-1}. These figures show that particles smaller than 10 μm will only reach the ground with reduced wind speed. Fallout is the most important deposition process during 'Harmattan' events.

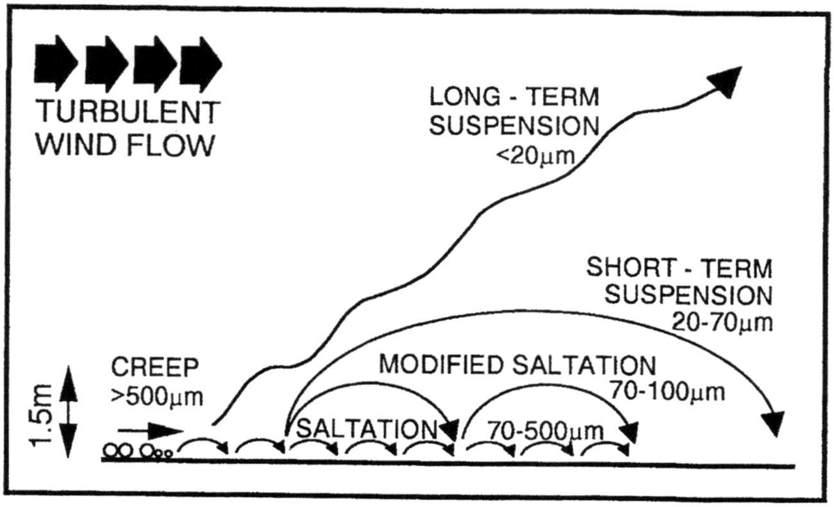

Fig. 2.1–2. Modes of particle transport by wind. (*Source*: Pye 1987, modified).

A second process is the capturing of particles by rough, moist, or electrically charged surfaces. This process is, in principle, well-understood but cannot be quantified because the availability of those surfaces and the probability of certain particles meeting the surface is not yet measurable with sufficient accuracy. Nevertheless, it can be concluded that rough soil surfaces like stony, ploughed, or ridged surfaces and especially surfaces of living or dead vegetation, are very attractive for dust particles. Some older plants grow on hummocks because they are continuously catching aeolian sediments and collecting them underneath their canopy. Such a process could be demonstrated by the presence or absence of high amounts of millet crop residues on soil surfaces of Luvic Arenosols in southwest Niger. Bationo and Mokwunye (1991) and Bley (1990) reported that plots with crop residues had a surface about 0.6 m above plots without residues after a period of six years of continuous cropping. In this case, as in Chapter 3.3, one has to take into account that not only dust but especially fine sand from the next plot was caught and stabilized by the crop residues at the soil surface.

A third possibility for deposition is that particles may aggregate in the air and, therefore, reach a higher sinking velocity than before. This may be due to charging in the air, to collecting moisture at the dew point, or finally, to solid particles forming aggregates with precipitated salt crystals. In all these cases, the material settles down to the ground faster. The reason for the deposition is often not known because it occurs in the air at an unknown distance from the soil surface. Therefore, the process is neither well-analyzed nor understood.

Table 2.1–1. Dust deposition rates in the Sahel in different regions and during different time periods reported by different authors

Authors	Region	Sampling time	Deposition ($t\ km^{-2}\ a^{-1}$)
Maley (1980)	Chad	1966–1967	109
McTainsh and Walker (1982)	N-Nigeria	1976–1979	137–181
Drees *et al.* (1993)	SW-Niger	1987–1989	164–212
Herrmann *et al.* (1994a)	SW-Niger	1992–1994	62–186

Very important processes that return particles to the soil surface, or even into the soil, are rainout and washout during precipitation events. Water from the air condenses on all kinds of hydrophilic solid particles. With the addition of water, particles become heavier and heavier and finally rainfall starts (rainout). It is a well-known phenomenon that even after a relatively short summer rain the air is clear because all dust particles are washed out. Of all the dust which reaches the ground, the greater part is the washout fraction. This also applies to the huge loess deposits of glacial or desert fringe origin (Alaily 1984).

According to Yaalon (1987), loess deposits need to receive an annual load of about 200 $g\ m^{-2}$, equivalent to 2 $t\ ha^{-1}$, over a period of at least 10 000 years to be recognized as pure loess deposits and then loess-derived soils. Soils with lower deposition rates show properties of the material derived from the solid rocks or sediments locally present. Below 6 $g\ m^{-2}\ a^{-1}$ the dust import still affects soil properties but is morphologically no longer recognized in the soil profile. Data reported in recent years from the West African Sahel have indicated that between 60 to more than 200 $g\ m^{-2}$ are deposited annually (Table 2.1–1). This means they are well in the admixture range of soils and will affect soil morphology as well as soil properties.

The annual deposition rate is the sum of several dust deposition events, partly as rainout after convective storms and partly as fallout with 'Harmattan' events (Plate 5). Therefore, the frequency of dust storms may severely change the dust yield in a certain area. The general picture is a slight decrease of dust import from desert fringes towards savannah regions. The fact that deposition depends on single events also causes seasonal or casual fluctuations of dust properties. They may be explained by the different trajectories and different admixtures of local material during transport from the source to the sink. Nevertheless, one can assume that the dominant properties will be derived from the major source. Dust samples collected with dust traps (Plate 4) showed an obvious seasonal variability (Table 2.1–2). For example, the pH differed from 6.3 to 7.3 and the electric conductivity (EC) varied by a factor of 10.

Table 2.1–2. Seasonal variation of chemical dust properties at Sadoré, Niger in 1992

Month	pH (H_2O)	EC ($\mu S\ cm^{-1}$)	CEC ($cmol_c\ kg^{-1}$)	Water-soluble ($mg\ kg^{-1}$) Na	K	Ca	Mg
			'Harmattan' season				
February	7.3	576	207	846	506	828	158
April	7.2	355	167	211	183	808	98
			Rainy season				
May	7.0	52	70	52	55	163	33
June	6.9	56	133	14	24	114	21
August	6.3	107	182	39	63	136	15
			'Harmattan' season				
November	7.1	589	173	744	444	697	124
December	7.3	451	199	430	355	639	113

Less variability was observed with the total cation exchange capacity (CEC), whereas a hundredfold difference between certain water-soluble cations was observed.

Another unsolved question is the prediction of the net deposition rate of dust. All dust particles which fall out or rain out, may later be taken up by another dust storm and carried to a different place. The remobilization of dust already deposited is very difficult to estimate. Surfaces of loamy or clayey unvegetated or temporarily uncovered soils are highly susceptible to wind erosion of dust particles. On the other hand, soil surfaces covered with permanent vegetation, sufficient crop residues, coarse stones or boulders, or with a rough undulating surface are effective in trapping the aeolian dust. This can be seen in extreme desert soils of the *Hamada*, where the vesicular layer below the stones is almost entirely made up of desert loess (Gauer 1991).

Dust Properties

Dust is characterized by granulometry, mineralogy, and chemical properties. The physical properties of soils are mainly indirectly affected by imported dust due to the contribution of dust particles during the development of the soil structure.

Granulometry

A general problem connected with granulometric analysis of dust is that the grain-size fractions which are transported may differ from those analyzed.

Table 2.1–3. Grain-size distribution of dust samples (in %) reported by different authors

Authors	Region	$< 2\mu m$	$2–20\mu m$	$20–63\mu m$	$> 63\mu m$ $^{*}> 50\mu m$
McTainsh and Walker (1982)	N-Nigeria	10–30	60–80		< 10*
Moberg *et al.* (1991)	N-Nigeria	25	42	15	19
Whalley and Smith (1981)	N-Nigeria	85		15	
Drees *et al.* (1993)	SW-Niger	15	25	17	44*
Stahr *et al.* (1993)	SW-Niger	26	58	16	0

This is due to the fact that grains may be transported as aggregates and then analyzed after dispersion as single smaller grains. Depending on the analytical methods used, aggregates formed during and after sedimentation may not be disintegrated during the analytical procedure. Furthermore, the interpretation of results is difficult when fraction boundaries are not identical. Dust samples obtained in Niger and Nigeria showed a clay fraction averaging about 25% (Table 2.1–3). The maximum and median of the grain-size distribution was close to 20 μm.

Generally, particles coarser than 63 μm are infrequent, but contradictory results were reported by Drees *et al.* (1993). The comparison of the results is difficult because of different methods of analysis for grain-size and fraction boundaries. The fact that a second maximum was found in the fraction between 2 and 20 μm, shows that this fraction is prominent when dust transported from far is considered. Furthermore, Drees *et al.* (1993) reported that especially when convective storms with high wind speeds occurred, they collected relatively fine samples. This gives rise to the supposition that the coarser fractions found in the low yielding samples during 'Harmattan' events, which normally have lower wind speeds and should therefore transport finer particles, could be a result of either the analytical techniques used or the influence of birds sitting on the dust catchers. One can assume that the dust trapped in south-west Niger and northern Nigeria (Table 2.1–3) is mainly transported over long distances because the particles are finer than those observed in loess deposits of desert fringes and periglacial areas where the coarse silt fraction generally dominates (Maus and Stahr 1977). The aeolian addition to soils in the Sahel can already be identified by grain-size analysis because the local material (Table 2.1–8) is much coarser than the imported dust (Herrmann *et al.* 1996).

Table 2.1–4. Mineral composition of bulk dust samples reported by different authors

Authors	Region	Quartz	Feldspar	Mica	Kaolinite	Others
Adetunji and Ong (1980)	N-Nigeria	+++	–	–	+	smectite
McTainsh and Walker (1982)	N-Nigeria	+++	++	?	±	calcite
Moberg *et al.* (1991)	N-Nigeria	+++	+	+	±	chlorite
Whalley and Smith (1981)	N-Nigeria	+++	++	+	±	
Wilke *et al.* (1984)	N-Nigeria	+++	+	±	+	calcite
Drees *et al.* (1993)	SW-Niger	+++	±	±	+	calcite
Herrmann *et al.* (1993)	SW-Niger	+++	++	+	++	calcite, smectite
Tiessen *et al.* (1991)	N-Ghana	+++	+	+	+	

+++ dominant; ++ high; + medium; ± low; – not reported; ? uncertain

Mineralogical Composition

The granulometric composition provides information about the transport regime of dust and also about possible influences of dust on water and air regimes of the loaded soils. From the mineral composition, on the other hand, genetic information about sources and transport trajectories and also about the potential impact on plant nutrition can be obtained.

The mineral composition of the bulk dust samples from Niger, Nigeria, and Ghana during the last decade showed a clear dominance of quartz as a mineral resistant to physical and chemical weathering (Table 2.1–4). Quartz becomes dominant, the more the dust material is subjected to weathering and redistribution by erosion and deposition. Kaolinite was also found in all samples but, quantitatively, is not dominant. Feldspars were present in all samples but in very different quantities. Micas, as the fourth major rock-forming mineral in the area studied, showed relatively low but variable amounts. Also, a number of accessory minerals were recognized. These were calcite, chlorite, and smectite. Bio-mineral remnants like those from diatoms and sponges were mentioned by Whalley and Smith (1981). The fossils may indicate that some of the dust arriving in the West African Sahelian and Sudanian zones is of limnic or palaeo-limnic origin.

More than the mineralogy of the silt-size fraction, the clay-size fraction of dust is of further interest because of its high potential for carrying nutrients, particularly K and magnesium (Mg), and its higher surface area and, therefore, higher chemical reactivity. In areas of West Africa that are characterized by soils developed from sediments of the *Continental Terminal*, the soils are for the most part restricted to kaolinite as the dominant clay mineral. It is well-documented that kaolinite is also the dominant mineral in the clay fraction of dust (Table 2.1–5). However, in dust samples smectite, which cannot be

Table 2.1–5. Mineral composition of the clay fraction of dust samples (in %) reported by different authors

Authors	Region	Kaolinite	Smectite	Illite	Mixed-layer	Others
McTainsh and Walker (1982)	N-Nigeria	+++	+	+		
Moberg *et al.* (1991)	N-Nigeria	25	20	15		12 (chlorite) 7 (vermiculite)
Pickering (1980)	N-Nigeria	50–65	20–30	5–10		
Whalley and Smith (1981)	N-Nigeria	+		+	occasional	
Wilke *et al.* (1984)	N-Nigeria	+++		++	occasional	
Drees *et al.* (1993)	SW-Niger	+++	+	±		
Stahr and Herrmann (1995)*	SW-Niger	+++	++	+	occasional	

+++ dominant; ++ high; + medium; ± low; – not reported
*Results presented in this paper

detected in soils which are locally weathered, is found to be the second dominant clay mineral. Smectite is a clear tracer towards the arid zones of the Sahara. Additionally, illite was found in all the samples but sometimes in minor quantities. Some authors reported chloritic, vermiculitic, and mixed-layer minerals. This may be explained by the fact that some analyses were carried out with greater efforts to detect minor quantities of certain clay minerals. The high kaolinite content in some dust samples was probably due to the fact that these samples had local kaolinitic admixtures, whereas the assemblage of smectites and illites points to origins in playas, depressions, or former lakes as well as to wind eroded areas from crystalline rocks (Hammer 1994). The difference between the clay mineral fractions in the autogenic soil and in dust, gives information about qualitative and quantitative changes in soil properties at a certain site.

Chemical Properties

As reported for the mineralogical composition, the chemical composition of bulk dust samples collected in West Africa at different periods and locations shows no contradictions. There is a general sequence of the elements in decreasing order: silicium (Si) > aluminium (Al) > iron (Fe), calcium (Ca) > K, Mg > sodium (Na) (Table 2.1–6). The differences among the references

Table 2.1–6. Total element content of 'Harmattan' dust samples (in %) reported by different authors

Authors	Region	Na	K	Ca	Mg	Si	Al	Fe
Beavington and Cawse (1978)	N-Nigeria	0.5	1.2	5.0	1.4	n.d.	5.3	3.8
McTainsh and Walker (1982)	N-Nigeria	0.7	1.7	0.8	0.5	30.9	5.9	3.2
Moberg *et al.* (1991)[a]	N-Nigeria	0.0	1.2	0.1	1.0	25.8	11.6	5.0
Wilke *et al.* (1984)	N-Nigeria	1.1	2.4	2.6	0.5	26.7	6.4	3.8
Doyne *et al.* (1938)	S-Nigeria	0.6	1.3	3.8	1.2	23.1	5.5	3.0
Jones (1938)	S-Nigeria	0.6	1.3	3.8	n.d.	23.1	5.6	2.7
Stahr and Herrmann (1995)[b]	SW-Niger	0.5	1.6	1.9	0.8	29.5	8.2	5.1
Tiessen *et al.* (1991)	N-Ghana	0.1–1.5	0.5–3.0	0.7–3.9	0.4–1.6	n.d.	n.d.	n.d.

[a] Only clay fraction analyzed
[b] Results presented in this paper
n.d. = not determined

for a single element are great enough to conclude that different dust sources or different sampling periods were represented. This is of practical relevance because the elements Ca, Mg, and K are relevant as nutrients and, therefore, affect the dust impact in certain environments.

The data of Tiessen *et al.* (1991) showed that the whole range of values occurred at a certain site (Table 2.1–6). Therefore, the question of whether there were regional differences cannot ultimately be answered. However, there have been indications that different dust sources affect the western part of West Africa (Bergametti 1992). With the exception of the data reported by Tiessen *et al.* (1991), it seems that the Na concentration, which is mainly bound in albitic feldspars was rather stable, whereas the Ca concentration, bound mainly in carbonate and gypsum, showed higher fluctuations. Overall, the chemical properties showed a granitic or intermediate geochemical composition. The differences in the analytical data reported by Moberg *et al.* (1991), are clearly due to the fact that they have only analyzed the clay fraction which was dominated by kaolinite. The extremely high Al concentration was therefore reasonable.

As soon as short-term impacts of dust are in question, the more mobile fractions of elements have to be taken into account. As a first approach,

Table 2.1–7. Chemical properties of dust reported by different authors

Authors	Region	pH (H_2O)	CEC ($cmol_c\ kg^{-1}$)	Base saturation (%)	
Moberg *et al.* (1991)	N-Nigeria	7.0	360	98	
Drees *et al.* (1993)	SW-Niger	n.d.	60–220	100	
Herrmann *et al.* (1993)	SW-Niger	7.1	190	100	
		Exchangeable (mg kg^{-1})			
		Na	K	Ca	Mg
Moberg *et al.* (1991)	N-Nigeria	760	1460	4770	530
Drees *et al.* (1993)	SW-Niger	20–350	120–1570	1000–4800	120–490
		Water-soluble (mg kg^{-1})			
		Na	K	Ca	Mg
Drees *et al.* (1993)	SW-Niger	140–920	310–980	100–1000	40–210
Stahr and Herrmann (1995)*	SW-Niger	10–1030	5–510	15–1220	5–180

*Results presented in this paper
n.d. = not determined

one can state that 'Harmattan' dust does not carry much acidity, but it may buffer soil borne acidity. This is documented by the measured pH values as well as by the 100% base saturation (Table 2.1–7). The CEC is higher than expected for kaolinite, which reflects the influence of small-sized crystals and also of admixtures of illitic and smectitic clay. Again Moberg *et al.* (1991) reported the highest CEC values because these were obtained from the clay fraction. Overall, the concentration of water-soluble Na, K, Ca, and Mg can reach 2 g kg^{-1} of dust. Differences between the data reported by Drees *et al.* (1993) and results of this paper are mainly due to methodology rather than to natural differences. We collected samples monthly, therefore receiving particularly lower minimum values than Drees *et al.* (1993), who collected samples quarterly. Sodium was bound to a higher proportion in a water-soluble form, whereas K, Ca, and Mg are more important in the exchangeable fraction. The data reported by Moberg *et al.* (1991) concerning exchangeable cations are in agreement with the maximum values of Drees *et al.* (1993). This can be explained by the fact that Moberg sampled only during the 'Harmattan' period. Therefore, the maximum exchangeable cations were received, because during that period of the year no leaching before or during sedimentation had to be considered.

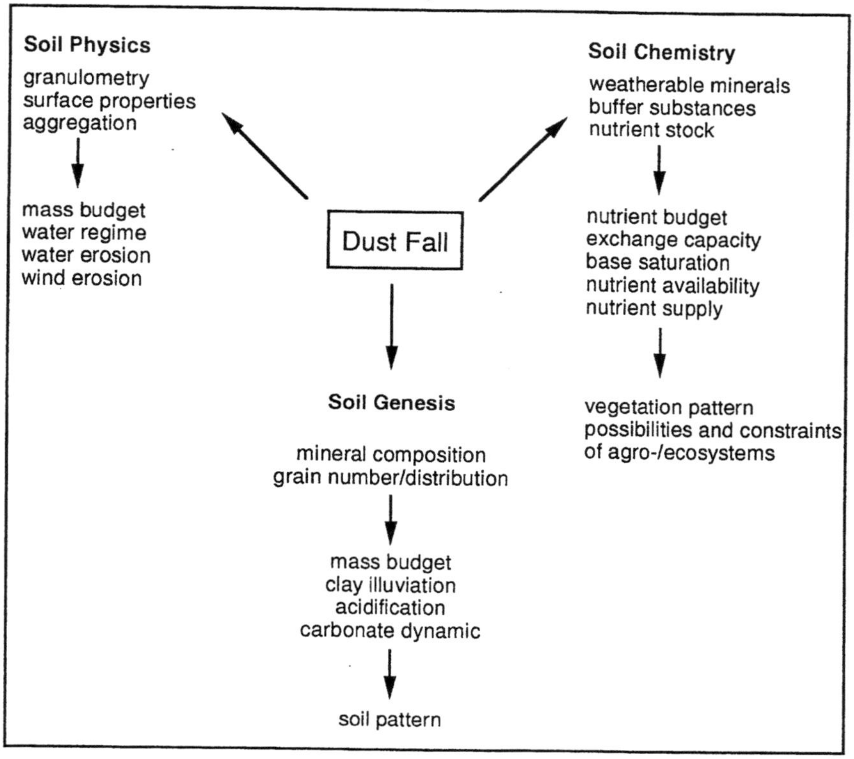

Fig. 2.1–3. Possible influence of dust on soil and site qualities.

Influence of Dust on Soil and Site Properties

The influence of dust on a certain soil depends on the amount of dust added and on the difference in the properties between the given soil and the added dust. According to Yaalon (1987) there is a full range of possibilities, from a soil and site development unaffected by dust to a soil totally derived from dust as in loess areas. Therefore, the dust influence on soil properties must be assessed from site to site. The type of influence is threefold (Fig. 2.1–3). Dust can affect soil genesis, soil physics, and soil chemistry. Soil genesis is affected as soon as dust addition has an impact which changes the direction or intensity of any soil forming process. Stahr *et al.* (1989) and Jahn (1995) gave examples where the addition of aeolian dust to the surroundings of the Mediterranean significantly changed the carbonate dynamics and also altered the texture development by quartz addition. Under these conditions, dust will prevent acidification. Concerning physical properties, dust import not only changes the particle size distribution, but even in small quantities severely changes the surface properties (Herrmann *et al.* 1993; Breuer 1994). As soon as the dust is incorporated into deeper soil layers, it influences the water and air regimes of the topsoil and strongly influences the erodibility of certain soils. The chemical impact of dust is mainly of interest when nutrients are considered. Non-nutrient elements may also influence the chemical status of the soils in question.

Dust as Soil Material

When dust is added to a certain soil, the mass budget is of initial interest (Table 2.1–1). A principal analytical problem is still the measurement of the net intake of dust at a certain site because any site may from event to event serve as a source or a sink area. The tendency of the arid part of a landscape serving more as a source area and the humid part as a sink area is a generalization. Specific events may change this situation totally. So far, all measurements and calculations have only been able to measure the gross intake. However, this need not be the maximum intake because surface roughness could lead to a higher input than observed over reference surfaces in dust samplers. Nevertheless one can assume that around the 500 mm rainfall isohyete, approximately 2 t ha^{-1} a^{-1} of dust are added to a vegetated soil surface in the West African sub-Saharan area. This is an equivalent of 0.15 mm topsoil (compare profiles documented by West *et al.* 1983 and Hammer 1994). This means the soil surface can rise 1 mm in seven years under optimum conditions, 0.1 m in 700 years, and 1 m in 7000 years. In the area under investigation, the net input is obviously less than calculated here, otherwise one would find loess soils (McTainsh 1984). On the other hand, eroded surfaces like lateritic ironcrusts show dust-derived covers, mainly in pockets, which can have a thickness from a few mm up to several cm (Plate 11). This observation makes it plausible that the rates observed with dust catchers do occur in the natural environment, at least in the last few decades. Studies on the effects of crop residues on surface soils (Chapter 3.3) underline this observation.

There is no doubt that the grain-size distribution of the dust differs from that of local soils in the sandy belt of the Sahel. From Table 2.1–8 it is obvious that Arenosols differ in their grain-size distribution significantly from the dust. Assuming that the silt fraction of the topsoil is primarily due to dust import, it can be concluded that only 5% of the topsoil is currently derived from dust. If this is true, only 25 to 30% of the actual clay fraction can be traced back to the dust input with the rest being locally formed. It is obvious from the depth function that the silt content in the topsoil is generally higher than in the subsoil, which shows the direct influence of dust. The dust addition, as long as it occurs as a separate layer, forms specific vesicular microhorizons at the soil surface causing desert pans (Yaalon 1987) or surface crusts (Herrmann *et al.* 1993). It is most likely that these materials are responsible for the properties of hard setting soils (Breuer 1994).

Table 2.1–8. Grain-size distribution (in %) of 'Harmattan' dust and topsoil at two sites (Arenosols) in southwest Niger in January 1992

Site	Sand	Silt	Clay
Ouallam			
Dust	0	68	32
Soil (0–0.1 m depth)	93	4	3
Sadoré			
Dust	0	74	26
Soil (0–0.1 m depth)	85	6	9

Effect on Water and Nutrient Budget

The added fine-grained dust material influences the pore-size distribution and the aggregation of a given soil, and therefore should have an influence on the water holding capacity. This simple assumption, however, has not yet been investigated in any particular case. The analytical proof is difficult because at a given site one would not find surfaces affected by dust and equivalent surfaces not affected by dust. However, the redistribution of water at the soil surface is affected when the added dust forms crusts. The crust hinders infiltration and therefore facilitates runoff (Casenave and Valentin 1989). A comparison of non-eroded and slightly eroded surfaces could be a tool for assessing the influence of dust.

The assessment of the influence on the nutrient budget is much easier. From data shown in Table 2.1–9, one can calculate that annually about 20 kg K ha^{-1}, 5 kg Mg ha^{-1}, 10 to 20 kg Ca ha^{-1} and finally a little less than 1 kg P $ha^{-1}$are added to farmers' fields in northern Nigeria and southwest Niger with the aeolian deposition. Deposition rates for the water-soluble and exchangeable element fractions can be much lower (Drees *et al.* 1993). The deposition is less than the amount needed by a growing crop. However, in agro-pastoral farming systems where the majority of the nutrient uptake by plants can be recycled with crop residues or farmyard manure, the addition of dust contributes to the maintenance of a good K and Mg supply.

In the long run, one may calculate that the total input is equal to the annual availability, because parts of the stock in the soil can always be mobilized by mineral weathering. Therefore after a longer period, a balance should develop where the addition marks the available amount. The fact that a rather high amount of K but less P is imported explains, at least in part, the severe P deficiency of Luvic Arenosols in southwest Niger (Bationo and Mokwunye 1991; Hafner 1992; Rebafka 1993). At the same time, the K nutrition of

Table 2.1–9. Annual nutrient deposition with dust reported by different authors

Authors	Region	Time period	Na	K	Ca	Mg	P
			Total (kg ha^{-1} a^{-1})				
McTainsh and Walker (1982)	N-Nigeria	Oct.–May (1976–1979)	6.9	16.8	7.9	5.0	1.0
Wilke *et al.* (1984)	N-Nigeria	Oct.–May (1979–1980)	10.9	24.2	25.8	5.0	1.0
Herrmann *et al.* (1993)	SW-Niger	Jan.–Dec. (1992)	2.1	14.6	9.3	5.2	1.0
Tiessen *et al.* (1991)	N-Ghana	Dec.–Feb. (1988–1989)	0.8	3.1	3.8	1.4	n.d.
			Water-soluble and exchangeable (kg ha^{-1} a^{-1})				
Drees *et al.* (1993)	SW-Niger	(1987–1989)	0.6	2.8	4.8	1.1	n.d.

n.d. = not determined

millet stands is in the range of normal supply. The fact that added K really causes a reasonable change in the K status of the soil is also ascertained by the findings of Hebel (1995). He reported that the exchangeable K in the soil is severely decreased by plant uptake throughout the vegetation period. However, at the beginning of the next rainy season (after the 'Harmattan' dust has been incorporated into the soil), the readily available amount of K has reached almost the same level as the year before.

The important role of K input from dust within the nutritional cycle can also be elucidated by the ratios between element concentrations in dust and in the topsoil as shown for three soils in southwest Niger (Table 2.1–10). The K concentrations had the highest dust/topsoil ratio followed by the ratio for the CEC and the Mg concentration. The ratio for P is lower and part of the P input with dust may result from flying insects that fall into the dust catchers. Besides K, Ca is of great relevance. Although the presence of Ca in topsoils cannot be detected by this method, the obvious additions by the dust maintain the pH, at least at the soil surface (Table 2.1–10).

Conclusions

Dust transport is of global importance for soils, but the Sahara and its surroundings are subject to far more dust mobilization and dust accretion than average. Dust consists mainly of silt- and clay-size particles which are transported over hundreds of kilometers in suspension in the air. In the Sahel, dust

Table 2.1–10. Chemical properties of dusts and soils as well as dust/soil ratios at three sites in southwest Niger in 1992

Site	pH H_2O	CEC ($cmol_c\ kg^{-1}$)	Na (%)	K (%)	Ca (%)	Mg (%)	P ($mg\ kg^{-1}$)
Ouallam							
Dust (January)	7.0	176	0.4	1.7	1.6	0.7	678
Dust (May)	6.9	187	0.1	1.0	0.4	0.3	584
Soil (0–0.1 m depth)	5.7	16	0.1	0.1	0.0	0.0	85
Ratio dust/soil		11	3	13	10	10	7
Sadoré							
Dust (January)	7.1	194	0.5	1.6	1.7	0.7	699
Dust (May)	6.6	144	0.1	1.1	0.4	0.3	560
Soil (0–0.1 m depth)	5.0	22	0.1	0.1	0.0	0.1	220
Ratio dust/soil		8	3	13	10	5	3
Tara							
Dust (December)	7.2	188	0.3	1.6	1.8	0.7	715
Dust (June)	6.9	99	0.2	1.1	0.4	0.3	460
Soil (0–0.2 m depth)	5.7	24	0.1	0.1	0.0	0.0	179
Ratio dust/soil		7	2	13	11	5	3

is mainly deposited by rainout after convective storms and by fallout with 'Harmattan' storms. The dust arriving in the Sahel is much finer than the bulk of the surface soils and specifically has higher mica, feldspar, and three-layer clay mineral contents than the surface soils. The soil water regime, although the database is poor, is mainly influenced at the soil surface by a change in the infiltration and runoff. The soil surface is altered by dust import mainly by the formation of surface crusts with vesicular pores which seal the surface against gas and water exchange. Potassium is the most important element imported with the dust. The current rate of K input is sufficient to maintain low input agricultural systems, whereas P input is too low to ameliorate the existing P deficiency.

Abstract

Dust particles are transported aeolian solids that influence the properties and fertility of soils at almost every place in the world. The Sahara, however, is regarded as the biggest source of soil-derived dust on the globe. The influence of dust is of particular significance in the nutrient poor sites of the West African sand belt. This chapter presents a review on the importance of dust as a material influencing soil properties and, especially, the soil fertility parameters. Some recent findings about the process of dust generation and accretion, as well as the properties of dust, are shown and discussed against the background of current knowledge. Special emphasis is given to the physical and chemical properties of West African soils and the influence of wind deposited dust on water and nutrient budgets. Potassium deposited with dust seems to be of great significance for the ecology of the area.

Résumé

Origine, déposition et composition de la poussière et conséquences sur les propriétés du sol et du site tout particulièrement pour les régions sémi-arides d'Afrique de l'Ouest

Les particules de poussière, matière solide transportée par le vent, influencent les propriétés et la fertilité des sols dans pratiquement toutes les parties du monde. Le Sahara est cependant considéré comme la plus grande source de poussière de la planète. L'influence de la poussière est d'un intérêt particulier pour les sols sableux d'Afrique de l'Ouest pauvres en éléments nutritifs. Ce chapitre traite de l'importance de la poussière en tant qu'intervenant dans les propriétés des sols et particulièrement leurs paramètres de fertilité. Des résultats récents concernant le processus de formation et l'accumulation de poussière, ainsi que ses propriétés sont présentés, discutés et comparés aux connaissances actuelles. L'on insistera sur les caractéristiques physiques et chimiques des sols ouest-africains et sur l'influence de l'apport de poussière dans les bilans des éléments nutritifs et d'eau. Le potassium transporté avec la poussière semble être d'une grande importance écologique pour la région.

References

Adetunji, J. and C.K. Ong. 1980. Qualitative analysis of the Harmattan haze by X-ray diffraction. *Atmospheric Environment* 14: 857–858.

Alaily, F. 1972. Entstehung von Decklehmen auf Lias α-Schichtflächen in Süddeutschland und deren Veränderung durch Bodenbildung. Diss. Universität Hohenheim, Stuttgart, Germany.

Alaily, F. 1984. Heterogene Ausgangsgesteine von Böden: Die Rekonstruierung und Bilanzierung von Böden aus heterogenen Ausgangsgesteinen. Landschaftsentwicklung und Umweltforschung. Schriftenreihe FB Landschaftsentwicklung, TU Berlin 25, Germany.

Bationo, A. and A.U. Mokwunye. 1991. Alleviating soil fertility constraints to increased crop production in West Africa. The experience in the Sahel. *Fertilizer Research* 29: 95–115.

Beavington, F. and P.A. Cawse. 1978. Comparative studies of trace elements in air particulate in northern Nigeria. *Science of the Total Environment* 10: 239–244.

Bergametti, G. 1992. Atmospheric cycle of desert dust. *Encyclopedia of Earth System Science* Vol. I: 171–182.

Bley, J. 1990. Experimentelle und modellanalytische Untersuchungen zum Wasser- und Nährstoffhaushalt von Perlhirse (*Pennisetum americanum* L.) im Südwest-Niger. Diss. Universität Hohenheim, Stuttgart, Germany.

Breuer, J. 1994. Tonmineralbestand von Böden aus unterschiedlichen Substraten im semiariden Kamerun. *Mitteilungen der Deutschen Bodenkundlichen Gesellschaft* 74: 333–336.

Casenave, A. and C. Valentine. 1989. Les états de surface de la zone sahélienne. Edition de l'ORSTOM, Paris, France.

Doyne, H.C., K.T. Hartley, and W.A. Watson. 1938. Soil types and manurial experiments in Nigeria. Third West African Agricultural Conference.

Drees, L.R., A. Manu, and L.P. Wilding. 1993. Characteristics of aeolian dusts in Niger, West Africa. *Geoderma* 59: 213–233.

Felix-Henningsen, P. 1992. Frühholozäne Feuchtzeitböden auf Altdünen der Ténéré und des Tchigai-Berglandes, Ost-Niger. *Würzburger Geographische Arbeiten* 84: 97–129.

Gauer, J. 1991. Bodenentwicklung und Bodengesellschaften vom Mittelmeer zur Qattara Depression in Nordwestägypten. Berliner Geowissenschaftliche Abhandlungen (A) 136, Berlin, Germany.

Gomes, L., G. Bergametti, G. Coudé-Gaussen, and P. Rognon. 1990. Submicron desert dusts: A sandblasting process. *Journal of Geophysical Research* 95: 13927–13935.

Hafner, H. 1992. Effect of organic and inorganic fertilizer application on growth and mineral nutrient uptake of pearl millet (*Pennisetum glaucum* L.) and groundnut (*Arachis hypogaea* L.) in an acid sandy soil of Niger. Diss. Universität Hohenheim, Verlag Ulrich E. Grauer, Stuttgart, Germany.

Hammer, R. 1994. Bodensequenzen und Standorteigenschaften im Südwest-Niger/Westafrika. Hohenheimer Bodenkundliche Hefte 22, Stuttgart, Germany.

Hebel, A. 1995. Der Einfluß der organischen Substanz auf die räumliche und zeitliche Variabilität des Perlhirse-Wachstums auf Luvic Arenosolen des Sahel (Fallstudie: ICRISAT in Sadoré/Niger). Hohenheimer Bodenkundliche Hefte 24, Stuttgart, Germany.

Herrmann, L., K.E. Bleich, and K. Stahr. 1993. Veränderung eines landwirtschaftlich genutzten Standortes in Niger/Westafrika durch Staubeintrag. *Mitteilungen der Deutschen Bodenkundlichen Gesellschaft* 72: 927–930.

Herrmann, L., K.E. Bleich, and K. Stahr. 1994a. Influence des dépôts éoliens sur la restauration de la fertilité des sols sous végétation en friche au Niger, Afrique de l'Ouest. *Réseau Erosion Bulletin* 14: 74–81.

Herrmann, L., B. Sponholz, and K. Stahr. 1994b. Quellregionen für den Harmattan-Staub in Westafrika: Ein mineralogischer und geochemischer Ansatz. *Mitteilungen der Deutschen Bodenkundlichen Gesellschaft* 74: 367–370.

Herrmann, L., K. Stahr, and M.V.K. Sivakumar. 1996. Dust deposition on soils of south-west Niger. *In* B. Buerkert, B.E. Allison, and M. von Oppen (eds.) Wind erosion in West Africa: The problem and its control. Proceedings International Symposium, Hohenheim, Germany, 5–7 Dec. 1994, Margraf Verlag, Weikersheim, Germany.

Jahn, R. 1995. Ausmaß äolischer Einträge in circumsaharischen Böden und ihre Auswirkungen auf Bodenentwicklung und Standortseigenschaften. Hohenheimer Bodenkundliche Hefte 23, Stuttgart, Germany.

Jones, B. 1938. Dessication and the West African colonies. *Geographical Journal* 91: 401–423.

Mainguet, M. 1991. Desertification. Natural background and human mismanagement. Springer Series in Physical Environment Vol. 9, Springer-Verlag, Berlin, Heidelberg, Germany.

Maley, J. 1980. Les changements climatiques de la fin du tertiaire en Afrique: Leur conséquence sur l'apparition du Sahara et sa végétation. pp. 63–86. *In* M.A.J. Williams and H. Faure (eds.) The Sahara and the Nile. Balkema, Rotterdam, The Netherlands.

Maus, H. and K. Stahr. 1977. Auftreten und Verbreitung von Lößbeimengungen in periglazialen Schuttdecken des Schwarzwaldwestabfalls. *Catena* 3: 369–386.

McTainsh, G.H. 1984. The nature and origin of the aeolian mantle of central northern Nigeria. *Geoderma* 33: 13–37.

McTainsh, G.H. and P.H. Walker. 1982. Nature and distribution of Harmattan dust. *Zeitschrift für Geomorphologie N.F.* 26: 417–435.

Moberg, J.P., I.E. Esu, and W.B. Malgwi. 1991. Characteristics and constituent composition of Harmattan dust falling in northern Nigeria. *Geoderma* 48: 73–81.

Morales, C. (ed.) 1979. Saharan dust - mobilization, transport, deposition. Scope 14, John Wiley and Sons, New York.

Pickering, J.G. 1980. X-ray diffraction analyses of Harmattan dust samples. Technical Memo, CSIRO Division of Soils 26.

Pye, K. 1987. Aeolian dust and dust deposition. Academic Press, London, U.K.

Rebafka, F.-P. 1993. Deficiency of phosphorus and molybdenum as major growth limiting factors of pearl millet and groundnut on an acid sandy soil in Niger, West Africa. Hohenheimer Bodenkundliche Hefte 9, Stuttgart, Germany.

Singer, A. 1986. Aluminosilicate clay mineral changes and soil formation in the hot-dry environment. pp. 1071–1080. Transactions of the XIII. Congress of ISSS, Hamburg, Germany.

Stahr, K., R., Jahn, A. Huth, and J. Gauer. 1989. Influence of aeolian sedimentation in soil formation in Egypt and Canary Island deserts. *Catena Supplements* 14: 127–144.

Stahr, K., K.E. Bleich, A. Hebel, L. Herrmann, C. Renard, and M.V.K. Sivakumar. 1993. The influence of organic matter and dust deposition on site characteristics and their microvariability. pp. 25–58. *In* University of Hohenheim (ed.) Standortgemäße Landwirtschaft in Westafrika – SFB308. Arbeits- und Ergebnisbericht (Zwischenbericht 1991–1993) (internal report).

Tiessen, H., H.-K. Hauffe, and A.R. Mermut. 1991. Deposition of Harmattan dust and its influence on base saturation of soils in northern Ghana. *Geoderma* 49: 285–299.

Völkel, J. 1988. Zum jungquatären Klimawandel im saharischen und sahelischen Ost-Niger aus bodenkundlicher Sicht. *Würzburger Geographische Arbeiten* 69: 255–276.

West, L.T., L.P. Wilding, J.K. Landeck, and F.G. Calhoun. 1984. Soil survey of the ICRISAT Sahelian Centre. Texas A&M University, TX.

Whalley, W.B. and B.J. Smith. 1981. Mineral content of Harmattan dust from northern Nigeria examined by scanning electron microscopy. *Journal of Arid Environments* 4: 21–29.

Wilke, B.M., B.J. Duke, and W.L.O. Jimoh. 1984. Mineralogy and chemistry of Harmattan dust in northern Nigeria. *Catena* 11: 91–96.

Wilson, S.J. and R.U. Cooke. 1980. Wind erosion. pp. 217–251. *In* M.J. Kirkby and R.P.C. Morgan (eds.) Soil erosion. John Wiley and Sons, London, U.K.

Yaalon, D.H. 1987. Saharan dust and desert loess: Effect on surrounding soils. *Journal of African Earth Sciences* 6: 569–571.

Yaalon, D.H. and E. Ganor. 1973. The influence of dust on soils during the quaternary. *Soil Science* 116: 146–155.

2.2 Wind Erosion and Wind Erosion Effects on the Growth and Development of Pearl Millet

Ludger Eltrop, Bruce E. Allison, and Karlheinz Michels

Introduction

Globally, wind erosion damage occurs on 430 million hectares, which is 8% of the dryland area (UNEP 1992). Wind erosion in the West African Sahel is a well-known phenomenon (e.g. Mainguet and Chemin 1977; Michels 1994). In the southern Sahelian zone, rainfall at the beginning of the rainy season is often preceded by convective storms that last from 15 to 60 min (Chapter 1.1). Typically, these storms transport large amounts of soil by saltation (Plate 1).

Sandstorms and wind-blown soil can damage plants either directly by causing leaf abrasion (Skidmore 1966) or covering the seedlings with soil (Fryrear and Downes 1975); or indirectly by decreasing photosynthesis (Armbrust 1982) and inhibiting plant development and growth (Armbrust 1984). However, the extent of plant damage depends largely upon wind speed; the amount, size, shape, and density of the abrasive material (Zobeck and Fryrear 1986a, 1986b); the frequency and duration of sandstorms; and the environmental conditions after exposure (Brenner 1991). Also, plant species (Fryrear and Downes 1975), plant age (Armbrust 1984), and plant shape markedly influence the extent of wind erosion damage. Consequently, predictions of plant damage due to wind erosion are difficult.

Strong winds may also cause a significant loss of the fertile topsoil, particularly of the sandy soils in West Africa (Raulin 1965; Chapters 3.3 and 3.4). Wendt (1984) reported a considerable loss of the fine soil particle fraction on the sandy soils in Niger, which caused a significant decrease in soil fertility and crop productivity. Dregne (1990) found a 50% reduction in soil productivity due to wind erosion for parts of southeast Tunisia. However, it is not clear whether long-term effects of wind erosion, such as the loss of soil fertility, exceed the short-term effects, such as abrasion damage and coverage of plants with soil. In contrast to the damaging effect of wind erosion to plants,

B. Buerkert, B. E. Allison and M. von Oppen (eds.), Wind Erosion in Niger, 67–83.

deposited wind-blown 'Harmattan' dust in the dry season may increase the nutrient status of the soil by providing mineral elements such as potassium, calcium, and magnesium (Drees *et al.* 1993; Herrmann *et al.* 1994; reviewed in Chapter 2.1).

Large areas of the arable land in the Sudano-Sahelian zone are cropped with pearl millet (*Pennisetum glaucum* (L.) R. Br.), usually in association with legumes like cowpea (*Vigna unguiculata* (L.) Walp.) (Chapter 1.3). In the traditional cropping systems in Niger, millet is sown with the first rains in May or June. Due to the strong sandstorms and dry spells after sowing (Chapter 1.1), millet establishment is often poor and large parts of the crop are covered with soil. This often makes replanting necessary (McIntire and Fussell 1989).

The extent to which local farmers recognize wind erosion as an important factor in plant productivity differs. Bargel *et al.* (1990) and Taylor-Powell *et al.* (1991) reported that peasant farmers in Niger considered crop damage due to wind erosion one of the major constraints in millet production, whereas in another study by Feil and Lamers (Chapter 4.3), wind erosion was not regarded as a major problem for millet growth.

The extent of wind erosion, determined as soil flux, and the effect of wind and wind-blown soil on the establishment and growth of pearl millet was investigated by Michels *et al.* (1993, 1995) in a field study at the ICRISAT Sahelian Center (Niger) and in wind tunnel experiments at the USDA Wind Erosion Laboratory, Manhattan, Kansas (USA). A summary of the results is given in this chapter.

Wind Erosion during the 1990 Growing Season and Its Effect on Millet Establishment

Soil Flux

During the 1990 growing season, 21 sandstorms were counted over a period of approximately three months. They occurred at irregular intervals and differed in duration and strength (Fig. 2.2–1). The extent of the sandstorms was determined by measuring the amount of wind-blown soil using the BSNE soil samplers (Fryrear 1986a; Plate 12) placed in the middle of a millet field. During the sandstorms, varying amounts of wind-blown soil were collected in the soil samplers (Fig. 2.2–1). During one single sandstorm on 6 July, 384 kg soil m^{-2} sampler opening were collected. However, observations indicated that most of the wind-blown soil was redistributed within a limited area and that the soil flux ended a few minutes after the onset of rainfall. As the height above ground level increased, the amount of wind-blown soil decreased exponentially and most of the soil (88% of total) was captured at 0.1 m height above ground level (Fig. 2.2–2). The annual amount of wind-blown soil at 0.1 m height amounted to 1261 kg m^{-2} sampler opening.

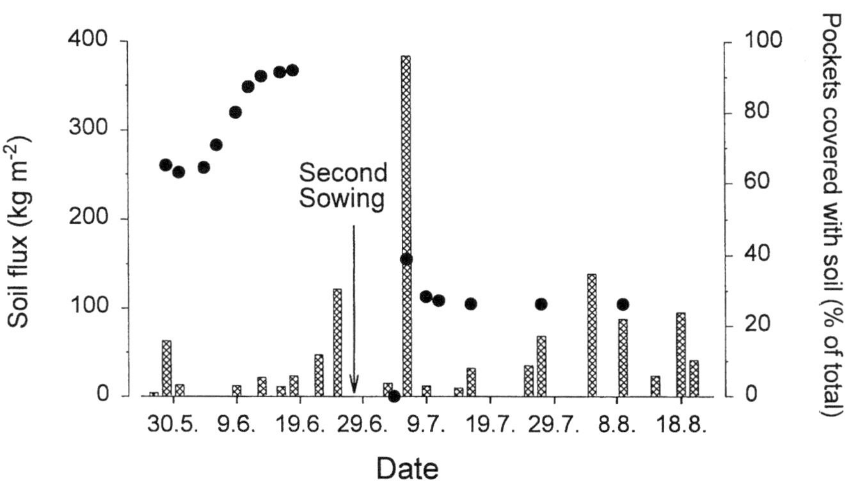

Fig. 2.2–1. Wind erosion events (sandstorms) and coverage of millet pockets with soil during the 1990 growing season at the ICRISAT Sahelian Center, Sadoré (Niger). Vertical bars represent the amount of soil captured in BSNE soil samplers at 0.1 m height above ground level. Black dots show the percentage of millet pockets totally covered with soil. Millet was sown a second time because 30 days after emergence more than 90% of the plants were covered with soil.

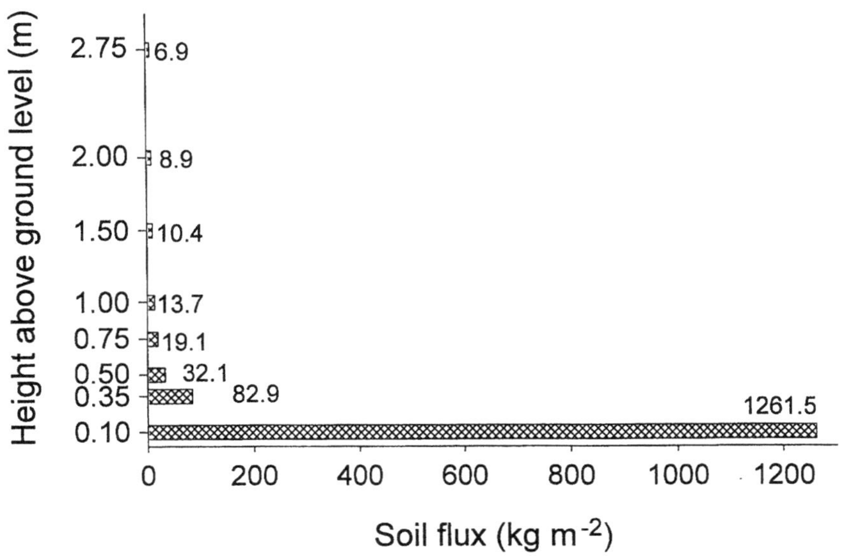

Fig. 2.2–2. Total amount of soil captured in BSNE soil samplers during sandstorms at different heights above ground level in the growing season 1990.

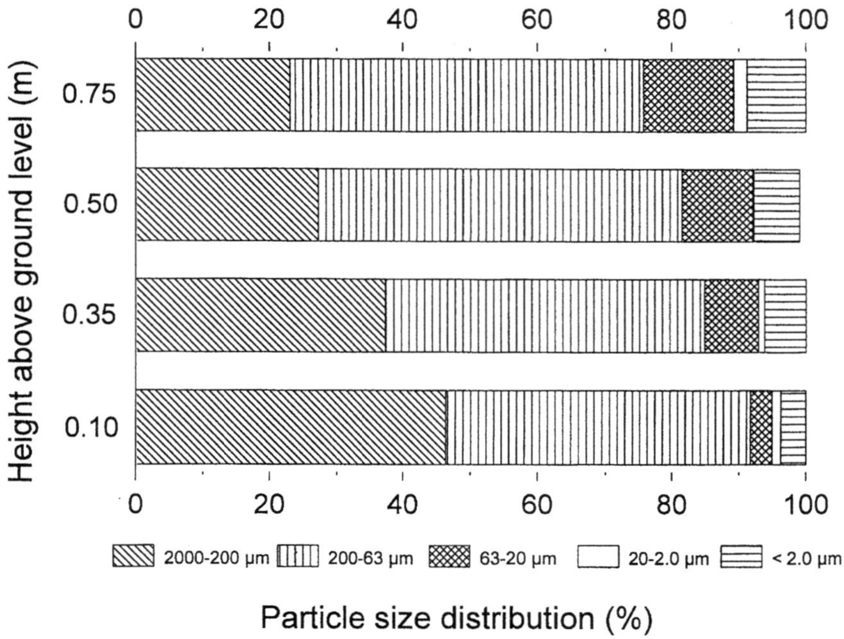

Fig. 2.2–3. Soil particle size distribution of aeolian material captured in the middle of a pearl millet field in BSNE soil samplers at different heights above ground level. Data are means of the sampling years 1990 to 1993.

The extent of soil flux during the 1990 growing season (Fig. 2.2–1) was comparable to that of 1991 and 1992 (Chapter 3.4). However, when compared with other semi-arid regions, the amounts of wind-blown soil at Sadoré (Niger) were relatively low. On a bare field in Big Spring, Texas (USA), 4267 kg soil m^{-2} sampler opening were collected at 0.15 m height over a period of 100 days (Zobeck and Fryrear 1986b). This difference between regions may not only be due to different environmental conditions (climate, vegetative cover) and different types of soil, but also to the duration of the sandstorms. In the semi-arid region of Big Spring (USA), the sandstorms lasted more than 10 h, much longer than those observed at the ICRISAT Sahelian Center (Niger).

As the height above ground level increased, the particle size distribution of the wind-blown material changed significantly. At 0.1 m height, the larger size particles (2000–200 μm) accounted for more than 46% of the material, whereas at 0.75 m height it accounted for only about 23% (Fig. 2.2–3). In contrast, the proportion of the medium size particles (20–200 μm) increased from 0.1 to 0.75 m height from 48 to 66%. The amount of soil particles smaller than 2 μm increased slightly from 3.8% at 0.1 m height to 8.8% at 0.75 m height.

The extensive soil flux during sandstorms (Fig. 2.2–2) may greatly contribute to soil erosion on the sandy soils of the Sahel. Buerkert *et al.* (Chapter 3.3) reported that due to combined wind and water erosion, 140 t ha^{-1} of soil were lost from bare plots between late May and late July 1994. On the other hand, the application of millet crop residues at a rate of 2000 kg ha^{-1} reduced soil flux significantly (Chapters 3.3 and 3.4), and topographical measurements of the soil surface revealed that 180 t ha^{-1} soil were deposited within three months (Chapter 3.3). Geiger *et al.* (1992) observed an increase of 0.15 m in the soil surface level after mulching with millet residues over a period of five years. The authors attributed this change, in part, to the use of millet crop residues which trapped wind-blown material and protected soil from erosion. Additionally, as described in the following chapters, natural savannah vegetation or planted windbreaks may also contribute to a reduction in wind speed and soil flux (Chapters 3.1 and 3.2). On windbreak plots with 10 m wide strips of the perennial grass *Andropogon gayanus*, 2000 t ha^{-1} of soil were trapped within three years (Renard and Vandenbeldt 1990).

In order to estimate the effect of wind erosion on soil degradation or rebuilding, a qualitative evaluation of gains and losses in terms of chemical and physical soil properties is necessary. Compared to an unprotected soil surface, wind-blown soil captured at 0.1 m above ground level contained higher amounts of calcium and magnesium (Chapter 3.4). This suggests that fertile soil particles and considerable amounts of mineral nutrients are retranslocated during sandstorms. This re-translocation was associated with a loss of finer soil particles and organic matter (Chapter 3.4). Consequently, sandstorms will accelerate the decline in the productivity of the poor sandy soils of the Sahel if the soils remain unprotected.

Establishment of the Pearl Millet Crop

Sandstorms may directly or indirectly affect plant growth and crop productivity. To investigate these effects, a field experiment was conducted. Millet was sown on 24 May 1990 in the experimental field. Sowing pockets were spaced 1 by 1 m, according to farmers' sowing practice. For millet, the most obvious and damaging effect of the sandstorms was the coverage of sown pockets with wind-blown soil (Fig. 2.2–1). The coverage with soil, particularly at an early stage of development, decreased plant establishment and survival as well as the growth of millet. During the first three sandstorms after sowing in May, about 80 kg soil m^{-2} sampler opening were collected at 0.1 m height above ground level and 60% of all millet pockets were totally covered with soil. Another 20% of the pockets were partially covered (one to four plants visible). No regrowth was observed in the totally covered millet pockets, whereas in the partially covered pockets some plants recovered and resumed growth. During the following four sandstorms, 68 kg soil m^{-2} sampler opening were collected and more pockets were totally covered with soil. At the end of June, only 3% of the pockets monitored remained uncovered and more than 90% were totally covered. The millet field was considered completely damaged and was resown on 25 June. In most cases, the complete coverage with soil killed the millet seedlings. This may, in part, have been due to high temperatures in the soil layer covering the plants. At the beginning of the rainy season, soil temperatures exceeding 50°C may occur in the upper 0.01 m soil horizon (Chapter 3.3). Soil-covered millet seedlings may also suffer from drought stress or a combination of both heat and drought stress. Additionally, the development of surface crusts may inhibit or delay the regeneration of the covered seedling. The results indicate that the coverage of millet seedlings with soil should be considered as a major constraint to plant establishment in the Sudano-Sahelian zone.

An exceptionally strong sandstorm, compared to the sandstorms in May and June, preceded the first rain after the second sowing on 9 July. In this storm the soil flux at 0.1 m height above ground level was 384 kg m^{-2} sampler opening (Fig. 2.2–1). In contrast to the first planting, less than 40% of the emerged millet plants were covered with soil, probably because the storm did not occur during the first twelve days after sowing. Some plants in the totally covered pockets recovered after the storm and the degree of pocket coverage decreased to less than 30%. The percentage of millet pockets covered by soil did not change significantly with subsequent storms, although the amount of wind-blown soil was sometimes considerable.

The results presented indicate that the degree of plant damage was not determined by the extent of the sandstorm *per se* but by the stage of plant development at the time of the erosion event. During one single sandstorm of minor strength three days after emergence (DAE) (first sowing), more millet pockets were covered than during a strong sandstorm at a later stage of plant development (second sowing). Pearl millet should therefore be protected against wind erosion damage particularly during the early sensitive stage of plant development.

The traditional way of sowing pearl millet in pockets may also be a measure to reduce plant damage by wind erosion. This planting system may prevent the total coverage of the emerging millet plants with soil and may increase the capacity to recover and resume growth after the coverage with soil. On the other hand, by compressing the soil in each pocket, a small depression is created which exacerbates the accumulation of wind-blown soil in the sown pockets and, subsequently, the partial or complete coverage of the emerging millet seedlings with soil.

In order to reduce the probability of plant damage by wind erosion in the early growing season, it is crucial to improve plant establishment and growth during the first sensitive phase of development. Bationo *et al.* (1992) and Buerkert and Stern (1995) reported that the application of nitrogen and phosphorus fertilizer promoted early growth of pearl millet in Niger and increased the final plant density by 49%. Klaij and Hoegmoed (1993) reported

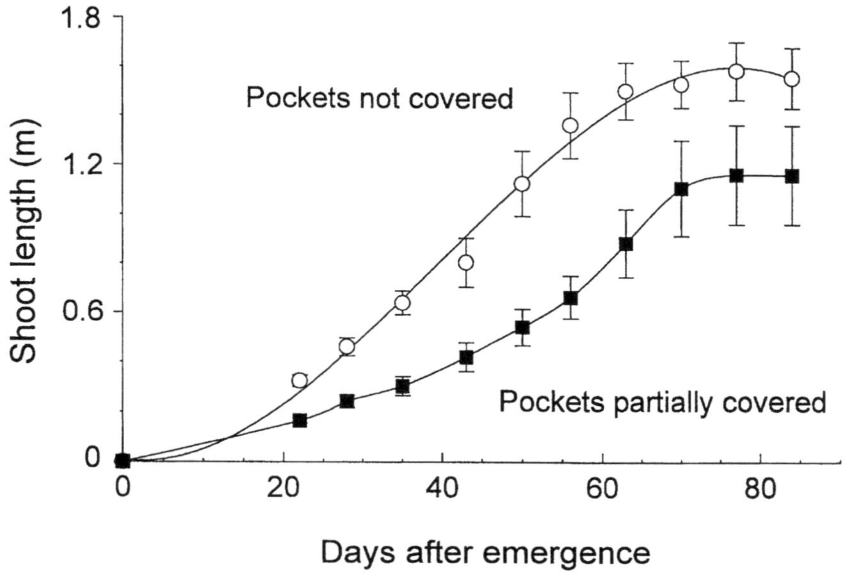

Fig. 2.2–4. Shoot length increment of millet plants (second sowing) not covered and partially covered with soil.

that 16 to 18 days after sowing, the millet stand was significantly increased by the application of mineral fertilizer (17%), by plowing (42%), and by ridging (54%). However, in Niger the possibilities and limitations of these means are well-known and have recently been discussed by Hafner (1992), Hafner *et al.* (1993), Klaij and Hoegmoed (1993), Rebafka (1993), and Bruentrup *et al.* (Chapter 4.2). Moreover, natural vegetation, planted trees, and crop residues may constitute protective means against wind erosion damage and improve crop growth (Part 3).

Growth and Development of Millet Plants after Coverage with Soil

In 1990, the main damaging effect of wind erosion on millet growth was the coverage of sown pockets with soil and the subsequent loss of the emerging plants. However, re-emerging plants from partially covered pockets along with pockets which were not covered may partly restore the millet yield. Therefore, the effects of wind erosion and particularly the effect of soil coverage on plant development was monitored in 40 randomly chosen quadrants (10 by 10 m) in the experimental field.

Visual observations of the surviving millet plants after the sandstorms showed that abrasion damage to leaves was minimal. Most crop damage was due to the coverage of seedlings with wind-blown soil. However, the surviving millet plants from pockets which were not covered and from partially covered

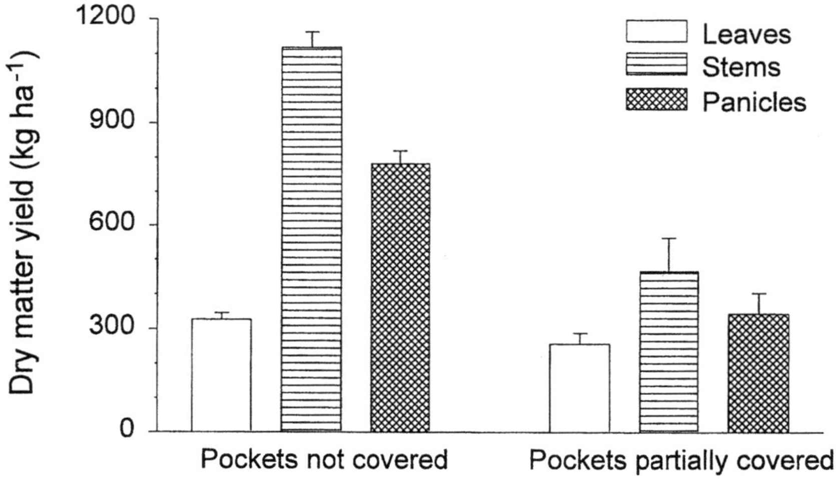

Fig. 2.2–5. Dry matter yield of leaves, stems and panicles of millet plants not covered (more than four plants visible per pocket) and partially covered (one to four plants visible per pocket) with soil.

pockets showed consistent differences in growth and phenological development during the growing season. Plant height was significantly reduced by the coverage of millet pockets with soil. Twenty-two DAE, the plants in the pockets which were not covered were significantly taller than the plants in the partially covered pockets and this difference was maintained throughout the growing season (Fig. 2.2–4). Ten weeks after emergence, the plants in the pockets not covered reached a maximum height of 1.6 m. The plants from partially covered pockets reached their maximum height two weeks later and were about 25% shorter.

At maturity, the dry matter of stems and panicles of the plants from partially covered pockets was significantly reduced compared to the plants from pockets not covered (Fig. 2.2–5). The decrease in dry matter production was 58% for stems and 56% for panicles. No significant difference in the dry matter production of leaves was observed between the plants from pockets not covered and partially covered.

In plants from partially covered pockets, panicle emergence began two weeks later than in plants which were not covered. Maximum panicle length occurred 13 weeks after emergence and was reduced by nearly 50% in the partially covered plants. Tiller and node number per plant were similarly reduced by the soil coverage of millet pockets (data not shown). Accordingly, the maximum leaf area index was much lower in the partially covered plants (0.35) compared to the plants not covered (0.85). In comparing the partially covered plants to the plants not covered, the number of leaves was also significantly reduced 28 DAE. However, the phenological differences became smaller later in the growing season (data not shown).

The growth reduction observed in partially covered millet plants indicates that besides the complete loss of plants due to soil coverage, wind erosion may also decrease the growth of the remaining plants. It appears that the delay in crop development may significantly reduce millet yields. Fryrear (1986b) observed that wind erosion caused a delay in plant growth from 10 to 28 days. Since the length of the rainy season and the amount of rainfall has decreased in many locations in Niger since 1965 (Chapter 1.1), many plants may not reach physiological maturity. On the other hand, smaller plants may survive longer dry periods while bigger plants suffer from drought stress due to higher water requirements.

Effects of Wind and Wind-blown Soil on Pearl Millet in Wind Tunnel Experiments

In order to determine the effects of wind erosion on growth and development of pearl millet under controlled environmental conditions (wind temperature 25°C; wind speed 8, 11, and 14 m s^{-1}; sand load of the wind 0, 8.3, 25.0, and 41.7 g m^{-1} s^{-1}), millet seedlings were exposed to wind or wind-blown sand in a wind tunnel. The plants were exposed to the different treatments for 15 min eight DAE.

Plant Survival and Leaf Damage

In all treatments, pearl millet survival was 100%. In contrast to millet, the survival rate of cotton, exposed to a sand flux of 50 g m^{-1} s^{-1} for 10 min, was 63% three DAE and 71% nine DAE (Fryrear 1971). A survival rate of 75% was found for seven vegetable species when exposed to 15 m s^{-1} wind and 50 g m^{-1} s^{-1} sand six DAE (Fryrear and Downes 1975). Thus, under wind tunnel conditions, pearl millet appears to be more resistant to sandblasting damage than other species.

The leaves of the plants exposed to wind alone were visibly damaged from flapping in the wind and whipping on the tunnel floor. In the sandblasting treatments, necrotic leaf edges, yellow spots, and a complete loss of leaves two to three days after the treatment were observed. Two days after the treatment, the viable leaf area was significantly reduced in the sandblasted millet plants compared to the plants exposed to wind alone (Fig. 2.2–6A). At a sand flux of 25 g m^{-1} s^{-1}, viable leaf area was reduced by 49% and at 41.7 g m^{-1} s^{-1} by 75%. No difference was observed between the wind treated plants and the unexposed control. However, due to the regrowth of leaves the difference in viable leaf area between sandblasted and wind treated plants decreased 13 days after exposure (DAExp.).

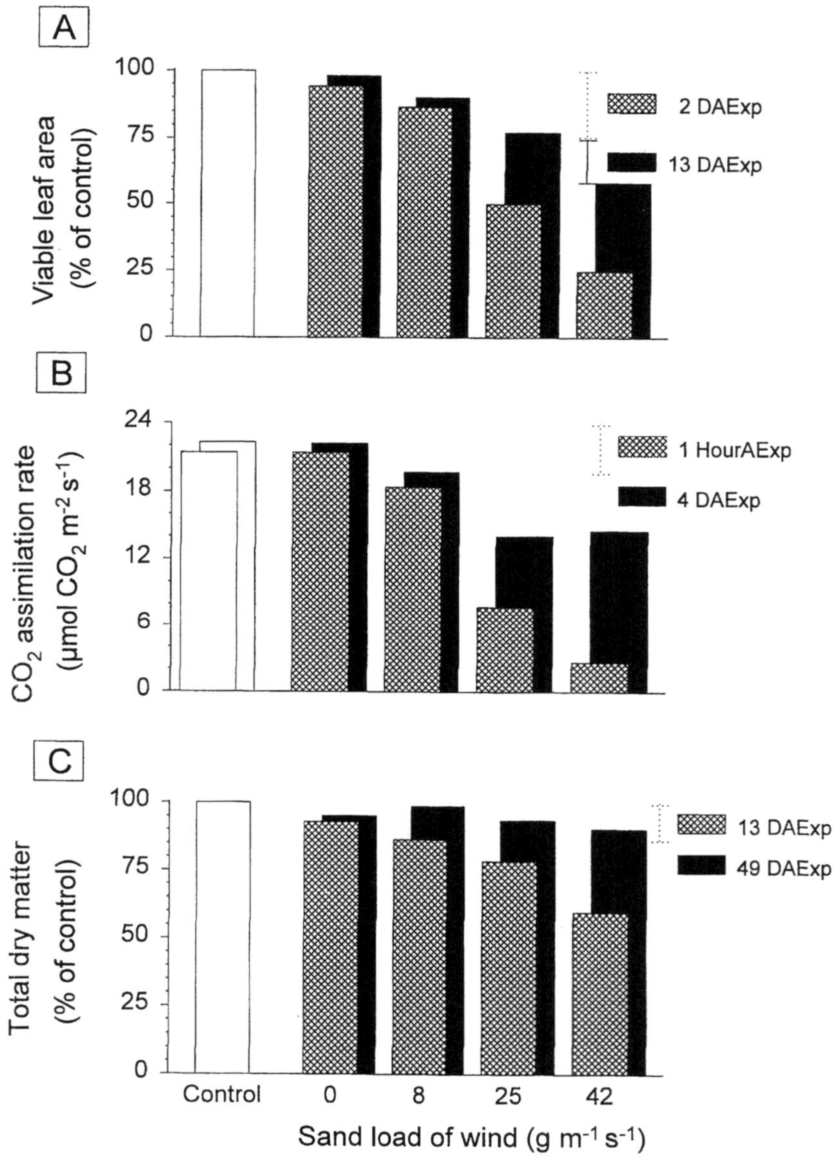

Fig. 2.2–6. Viable leaf area (A), photosynthesis (B), and above-ground dry matter (C) in millet plants exposed to wind and wind-plus-sand. Plants were subjected to the wind (11 m s^{-1}) and the wind-plus-sand treatment for 15 min eight days after emergence. Vertical lines represent the least significant difference (LSD). DAExp. = Days after exposure.

Damage to leaves is commonly observed after wind erosion events or wind and wind-plus-sand exposure (Greig *et al.* 1974; Precheur *et al.* 1978; Armbrust 1982). Generally, less plant damage has been found with plants exposed to wind alone than with combined exposure to wind-plus-sand. In tomato plants, exposure to wind alone caused no leaf damage, whereas extensive tissue destruction was observed when sand was added to the wind (Precheur *et al.* 1978). When exposed to wind alone (13.5 m s^{-1}) at the 9-leaf stage, viable leaf area in sorghum was reduced by 18%, whereas in the treatment with wind-blown soil the reduction was 60% (Armbrust 1982). In winter wheat, viable leaf area decreased by 63% with a wind-plus-sand treatment (Armbrust *et al.* 1974).

Due to the decrease in viable leaf area, the development of leaves in sandblasted plants may be delayed for several days. However, the delay is dependent on the developmental stage of the plants at the time of exposure. Fryrear (1971) reported that leaf area was significantly decreased and leaf development delayed by about 10 days in 3-day-old cotton seedlings exposed to wind-blown soil. When the plants were exposed 9 DAE, the delay in leaf development was between 10 and 17 days. However, Grace (1977) concluded that plants may tolerate and even benefit from the leaf damage and the delay in leaf development, as smaller plants are less susceptible to water shortages during and after the end of the wind erosion event. Only long-lasting exposure to wind may alter the wax structure of leaf surfaces (Grace 1977) and subsequently, change the loss of water by transpiration.

Photosynthesis

The treatment of millet with wind alone or with wind-plus-sand affected the CO_2 assimilation rate only immediately after exposure. In preliminary experiments, the assimilation rate of millet, subjected to wind alone (wind speed: 8 and 14 m s^{-1}) significantly decreased from 15.9 to 10.4 μmol CO_2 $m^{-2}s^{-1}$. This reduction was found only in the first hour after the treatment. Two days later, no significant difference was observed (data not shown). In further experiments, photosynthesis significantly decreased in sandblasted plants, compared to the control plants, particularly in the treatments with a sand flux of 25 g m^{-1} s^{-1} and higher (Fig. 2.2–6B), but only in the first hour after the treatment. Four days after exposure, photosynthesis was still slightly reduced in these plants but not to a significant degree (Fig. 2.2–6B). In the plants treated with wind alone, photosynthesis was not reduced significantly.

Effects on photosynthesis have been observed after the exposure of plants to wind and wind-plus-sand. In accordance with the data presented here, Precheur *et al.* (1978) reported a significant decrease in photosynthesis in tomato plants treated with wind and wind-plus-sand. In contrast in sorghum, photosynthesis increased by 48 to 85% after sandblasting (Armbrust 1982). A similar increase was found for winter wheat (Armbrust *et al.* 1974). Generally, the effects of a sandblasting treatment decreased with time after exposure. The degree and speed of recovery of the CO_2 assimilation rate may be one factor determining the resistance or susceptibility of plants to wind erosion damage.

The decrease in the short-term CO_2 assimilation rate in sandblasted plants may be caused either by leaf cell damage or by temporary drought stress. In tomato plants, considerable changes in leaf anatomy occurred after 20 min exposure to wind-plus-sand (Greig *et al.* 1974). Also, Precheur *et al.* (1978) reported leaf damage after wind-plus-sand exposure. In damaged leaves, the CO_2 assimilation rate may be decreased directly by the inhibition of the activity of the carboxylating enzymes or by the stimulation of leaf respiration (Armbrust 1982) or indirectly by short-term water deficits in the ruptured leaf tissues (Grace 1977).

Dry Matter Yield

Thirteen days after exposure, the above-ground pearl millet dry matter yield was significantly reduced in the 25 and 41.7 $g\,m^{-1}\,s^{-1}$ sandblasting treatments (Fig. 2.2–6C). In the plants treated with wind only, no difference was observed. However at final harvest, 49 days after exposure, there was no significant reduction in dry matter yield in the sandblasting treatments.

Considering the lower viable leaf area and the decreased CO_2 assimilation rate in the sandblasted plants shortly after exposure (Fig. 2.2–6A and B), it is not surprising that the dry matter yield also decreased a few days after the treatment (Fig. 2.2–6C). Reduced dry matter yield is often observed after a wind or wind-plus-sand treatment (Armbrust 1982). However, pearl millet showed a high capacity to recover from leaf damage and to regenerate new biomass, confirming similar results from Brenner (1991) in Niger. In contrast, cotton showed a higher sensitivity to wind and sandblasting damage. Fifty days after exposure to a wind-plus-sand treatment plant dry matter was still 30% lower than that of non-exposed control plants (Fryrear 1971). In wheat, sandblasting treatment caused a delay of up to ten days in the ripening of the grain (Woodruff 1956). In bean, pod yields decreased almost linearly with increasing wind speed, abrasive flux, and duration of exposure (Skidmore 1966).

The results of the wind tunnel experiments suggest that on a long-term basis, growth of pearl millet is not significantly affected by single wind erosion events. Millet will most likely survive a short sandblasting treatment as long as the plant damage after wind erosion is not aggravated by other stresses, such as high soil temperature or mechanical impedance due to the development of a surface crust, even if the damage occurs at an early stage of development. The results of the wind tunnel experiments show that the exposure to wind and wind-plus-sand may temporarily decrease the growth of millet and that the differences in viable leaf area and photosynthesis diminish a few days after the treatment.

Environmental conditions in the field are quite different from those in the wind tunnel. Wind erosion events occur more often during the growth period of a crop. They may last longer and carry more soil particles, thus, causing more damage to leaves, especially at early stages of growth. Also, plants recover from wind erosion damage much slower in the field than in the laboratory. Harsh climatic conditions, lack of water, and nutrient imbalances often limit the potential to regenerate damaged plant tissue (Armbrust *et al.* 1969). Therefore under field conditions, wind erosion may be more damaging to millet production than suggested by the data from the wind tunnel experiments.

Conclusions

The present study has shown that the covering of emerging millet seedlings or sown pockets with soil is the main wind erosion damage to pearl millet. The extent of plant damage primarily depends on the stage of plant development at the time of exposure. In order to reduce this damage, it is important to stimulate plant growth in the early phase of crop establishment. This may be achieved mainly by applying mineral fertilizers and crop residues to the field, as well as by ridging. However, mineral fertilizer and crop residues are of limited availability in West Africa, while ridging is not a common practice of Nigerien farmers.

Furthermore, it was shown that the partial coverage of millet plants with soil caused a significant decrease in growth and dry matter yield. However, the traditional way of sowing millet in seedling clusters may reduce the percentage of plants covered with wind-blown soil and increase the dry matter yield. The compression of soil after sowing may lead to an increased accumulation of soil in the sown pocket and may promote the formation of a surface crust which is detrimental to the growth and regeneration of millet after coverage with soil.

Due to the different environmental conditions, the results of the wind tunnel experiments may only be of limited use for predicting plant damage due to wind erosion under field conditions. Field studies are, therefore, indispensable in evaluating the effect of wind erosion on plant growth and productivity. However, the wind tunnel data may serve for calibrating wind erosion models such as the new Wind Erosion Prediction System (Hagen 1991; Bilbro 1992). They may also improve models like the Erosion Productivity Impact Calculator (Williams *et al.* 1990) and the Windbreak Economic Model (Kort *et al.* 1993).

In summary, the data suggest that in West Africa wind erosion constitutes a major constraint to millet growth and productivity. In order to reduce the damage to this crop, control measures against wind erosion should be installed on a broader scale. Among these measures surface covers like crop residues, windbreaks, and soil tillage have been shown to reduce the susceptibility of both soils and crops to wind erosion (Skidmore and Hagen 1977; Bilbro and Fryrear 1985; Fryrear 1987, 1989). These measures were tested at the ICRISAT Sahelian Center for their suitability in the Sahelian environment (Part 3).

Abstract

Strong convective storms at the beginning of the cropping season affect plant establishment in the Sahel. Therefore, both the extent of wind erosion and the effects of wind and wind-blown soil on the growth and development of pearl millet (*Pennisetum glaucum* (L.) R. Br.) were investigated under field conditions at the ICRISAT Sahelian Center, Sadoré (Niger) and in wind tunnel experiments. During the 1990 growing season, 21 sandstorms were counted in the area of the experimental millet field. These occurred at irregular intervals during the growing season and affected the millet plants at various stages of development. The extent of the sandstorms and their effect on millet establishment varied significantly from storm to storm. During one single sandstorm in July, 384 kg eroded soil m^{-2} sampler opening was collected in soil traps and 40% of the 11-day-old millet plants were totally covered with wind-blown soil.

In the wind tunnel experiments, 100% of the 8-day-old pearl millet seedlings survived the wind and the sandblasting exposure. Viable leaf area, photosynthesis and above-ground dry matter decreased significantly after sandblasting, but not after exposure to wind alone. However, the differences decreased after a few days and 49 days after exposure the dry matter yield of the sandblasted plants was not significantly different from the non-exposed plants. Pearl millet production in Niger may be affected by wind erosion only when severe sandstorms occur during an early sensitive stage of millet establishment and emerging plants become covered with wind-blown soil. After that stage the short sandstorms are not likely to cause significant wind erosion damage to pearl millet.

Résumé

L'érosion éolienne et ses effets sur la croissance et le développement du petit mil

De forts courants de convection au début de la saison des cultures affectent leur mise en place au Sahel. C'est pourquoi l'étendue de l'érosion éolienne, les effets du vent et de la terre transportée par ce dernier sur la croissance et le développement du petit mil (*Pennisetum glaucum* (L). R. Br.) ont été étudiés au champ au Centre Sahélien de l'ICRISAT, Sadoré (Niger) et grâce à des expériences en tunnel à vent. En 1990, durant la saison de croissance des cultures, 21 tempêtes de sable ont été dénombrées dans la zone expérimentale de plantation du mil. Ces tempêtes de sable se sont produites à intervalles réguliers pendant la période de croissance et ont affecté les plants de mil à divers stades de leur développement. La portée des tempêtes de sable et leur effet sur l'implantation du mil varient de façon significative d'une tempête à une autre. Durant une seule tempête de sable en juillet, 384 kg de sol érodé par mètre carré d'échantillon ouvert ont été trouvés dans les collecteurs et 40% des plants de mil âgés de 11 jours ont été totalement recouverts par ce sol transporté par le vent.

Dans les expériences menées dans le tunnel à vent, 100% des semis de petit mil âgé de 8 jours ont survécu au vent et à l'exposition aux rafales de sable. Les parties vitales des feuilles, l'activité photosynthétique et la quantité de matière sèche en surface décroissent de façon significative après avoir subi les rafales de sable, mais pas après une exposition au vent seul. Cependant, les différences s'amenuisent quelques jours après l'exposition, et 49 jours après celle-ci, les différences de production de matière sèche dans le cas des plants non exposés et ceux qui ont été soumis aux rafales de sable, ne sont pas significatives. La production de petit mil au Niger peut être affectée par l'érosion éolienne seulement si de fortes tempêtes de sable se produisent durant le tout premier stade sensible de son implantation et si les plants apparus sont recouverts de sol transporté par le vent. Après cette période de croissance, de brèves tempêtes de sable ne sont pas censées causer de grands dommages par érosion éolienne au petit mil.

References

Armbrust, D.V. 1982. Physiological responses to wind and sandblast damage by grain sorghum plants. *Agronomy Journal* 74: 133–135.

Armbrust, D.V. 1984. Wind and sandblast injury to field crops: Effect of plant age. *Agronomy Journal* 76: 991–993.

Armbrust, D.V., J.D. Dickerson, and J.K. Greig. 1969. Effect of soil moisture on the recovery of sandblasted tomato seedlings. *Journal American Society of Horticultural Science* 94(3): 214–217.

Armbrust, D.V., G.M. Paulsen, and R.Jr. Ellis. 1974. Physiological responses to wind- and sandblast-damaged winter wheat plants. *Agronomy Journal* 66: 421–423.

Bargel, G., I. Grimm, J. Haas, I. Harder, M. Noll, R. Schwarzmeier, and G. Strunden. 1990. Desertifikationsbekämpfung und soziale Organisation – Möglichkeiten und Grenzen der Implementierung technischer Maßnahmen zum Erosionsschutz im Niger. Schriftenreihe FB Internationale Agrarentwicklung, TU Berlin 133, Germany.

Bationo, A., C.B. Christianson, W.E. Baethgen, and A.U. Mokwunye. 1992. A farm-level evaluation of nitrogen and phosphorus fertilizer use and planting density for pearl millet production in Niger. *Fertilizer Research* 31: 175–184.

Bilbro, J.D. 1992. Sunflower dry matter production and plant structural relationships for wind erosion modeling. *Journal of Soil and Water Conservation* 47(2): 194–197.

Bilbro, J.D. and D.W. Fryrear. 1985. Effectiveness of residues from six crops for reducing wind erosion in a semi-arid region. *Journal of Soil Water Conservation* 40(4): 358–360.

Brenner, A.J. 1991. Tree-crop interactions within a Sahelian windbreak system. Ph.D. Diss. University of Edinburgh, Edinburgh, U.K.

Buerkert, A. and R.D. Stern. 1995. Effects of crop residue and phosphorus application on the spatial variability of non-destructively measured millet growth in the Sahel. *Experimental Agriculture* 31: 429–449.

Drees, L.R., A. Manu, and L.P. Wilding. 1993. Characteristics of aeolian dusts in Niger, West Africa. *Geoderma* 59: 213–233.

Dregne, H.E. 1990. Erosion and soil productivity in Africa. *Journal of Soil and Water Conservation* 45(4): 431–436.

Fryrear, D.W. 1971. Survival and growth of cotton plants damaged by windblown sand. *Agronomy Journal* 63: 638–642.

Fryrear, D.W. 1986a. A field dust sampler. *Journal of Soil and Water Conservation* 41(2): 117–120.

Fryrear, D.W. 1986b. Wind erosion and its effects on plant growth. *Annals of Arid Zones* 25(4): 245–254.

Fryrear, D.W. 1987. Ridging reduces wind damage to cotton. *Applied Agricultural Research* 1(5): 311–314.

Fryrear, D.W. 1989. Wind erosion: Mechanics, measurement, and control. pp. 117–123. *In* Soil, crop, and water management systems for rainfed agriculture in the Sudano-Sahelian zone. Proceedings International Workshop, ICRISAT, Niamey, Niger, 7–11 Jan. 1987, ICRISAT, Patancheru, A.P. 502324, India.

Fryrear, D.W. and J.D. Downes. 1975. Estimating seedling survival from wind erosion parameters. *Transactions of the ASAE* 18(5): 888–891.

Geiger, S.C., A. Manu, and A. Bationo. 1992. Changes in a sandy Sahelian soil following crop residue and fertilizer additions. *Soil Science Society of America Journal* 56: 172–177.

Grace, J. 1977. Plant response to wind. Academic Press, London, U.K.

Greig, J.G., N. Bokhari, D.V. Armbrust, and L.C. Anderson. 1974. Residual effects of wind and sandblast-damage on tomato plants at different stages of development. *Journal American Society of Horticultural Science* 99(6): 530–534.

Hafner, H. 1992. Effect of organic and inorganic fertilizer application on growth and mineral nutrient uptake of pearl millet (*Pennisetum glaucum* L.) and groundnut (*Arachis hypogea* L.) in an acid sandy soil of Niger. Diss. Universität Hohenheim, Verlag Ulrich E. Grauer, Stuttgart, Germany.

Hafner, H., E. George, A. Bationo, and H. Marschner. 1993. Effect of crop residues on root growth and nutrient acquisition of pearl millet in an acid sandy soil in Niger. *Plant and Soil* 150: 117–127.

Hagen, L.J. 1991. A wind erosion prediction system to meet user needs. *Journal of Soil and Water Conservation* 46(2): 106–111.

Herrmann, L., K.E. Bleich, and K. Stahr. 1994. Influence des dépôts éoliens sur la restauration de la fertilité des sols sous végétation en friche au Niger, Afrique de l'Ouest. *Bulletin Erosion* 14: 367–370.

Klaij, M.C. and W.B. Hoogmoed. 1993. Soil management for crop production in the West African Sahel. II. Emergence, establishment, and yield of pearl millet. *Soil and Tillage Research* 25: 302–315.

Kort, J., L. Brandle, and B. Johnson. 1993. Economic evaluation of field windbreaks, BECON and WBINT. A user's manual. PFRA Shelterbelt Centre, Indian Head, Saskatchewan, Canada.

Mainguet, M. and M.C. Chemin. 1977. Les marques de l'érosion éolienne dans le Sahel du Niger d'après les images satellites et les photographies aériennes. pp. 139–149. *In* Colloque pédologie télédétection, AISS, Rome, Italy.

McIntire, J. and L.K. Fussell. 1989. On-farm experiments with millet in Niger: Crop establishment, yield loss factors and economic analysis. *Experimental Agriculture* 25: 217–233.

Michels, K. 1994. Wind erosion in the Southern Sahelian zone. Diss. Universität Hohenheim, Verlag Ulrich E. Grauer, Stuttgart, Germany.

Michels, K., M.V.K. Sivakumar, and B.E. Allison. 1993. Wind erosion in the Southern Sahelian Zone and induced constraints to pearl millet production. *Agricultural and Forest Meteorology* 67: 65–77.

Michels, K., D.V. Armbrust, B.E. Allison, and M.V.K. Sivakumar. 1995. Wind and wind-blown sand damage to pearl millet. *Agronomy Journal* 87(4): 620–626.

Precheur, R., J.K. Greig, and D.V. Armbrust. 1978. The effects of wind and wind-plus-sand on tomato plants. *Journal American Society of Horticultural Science* 103(3): 351–355.

Raulin, H. 1965. Techniques et pases socio-économiques des sociétés rurales nigériennes. Etudes nigériennes 12, IFAN-CNRS, Niamey, Niger.

Rebafka, F.-P. 1993. Deficiency of phosphorus and molybdenum as major growth limiting factors of pearl millet and groundnut on an acid sandy soil in Niger, West Africa. Hohenheimer Bodenkundliche Hefte 9, Stuttgart, Germany.

Renard, C. and R.J. Vandenbeldt. 1990. Bordures d'*Andropogon gayanus* Kunth. comme moyen de lutte contre l'érosion éolienne au Sahel. *Agronomie Tropicale* 45(3): 227–231.

Skidmore, E.L. 1966. Wind and sandblast injury to seedling green beans. *Agronomy Journal* 58: 311–315.

Skidmore, E.L. and L.J. Hagen. 1977. Reducing wind erosion with barriers. *Transactions of the ASAE* 20(5): 911–915.

Taylor-Powell, E., A. Manu, S.C. Geiger, M. Ouattara, and A.S.R. Juo. 1991. Integrated management of agricultural watersheds: Land tenure and indigenous knowledge of soil and crop management. TropSoils Bulletin No. 91–04, Soil Management Collaborative Research Support Program, North Carolina State University, Raleigh, NC.

UNEP. 1992. Atlas of desertification. United Nations Environment Programme (ed.) Edward Arnold, Dunton Green, U.K.

Wendt, C. 1984. Soil management. p. 15. *In* L.T. West, L.P. Wilding, J.K. Landeck, and F.G. Calhoun. Soil survey of the ICRISAT Sahelian Center, Niger, West Africa. Soil and Crop Sciences Department/TropSoils, Texas A&M University System, College Station, TX.

Williams, J.R., P.T. Dyke, W.W. Fuchs, V.W. Benson, O.W. Rice, and E.D. Taylor. 1990. Erosion/productivity impact calculator: 2. User manual. *In* A.N. Sharply and J.R. Williams (eds.) U.S. Department of Agriculture Technical Bulletin No. 1768.

Woodruff, N.P. 1956. Wind-blown soil abrasive injuries to winter wheat plants. *Agronomy Journal* 48: 499–504.

Zobeck, T.M. and D.W. Fryrear 1986a. Chemical and physical characteristics of wind-blown sediment I. Quantities and physical characteristics. *Transactions of the ASAE* 29(4): 1032–1036.

Zobeck, T.M. and D.W. Fryrear. 1986b. Chemical and physical characteristics of wind-blown sediment II. Chemical characteristics and total soil and nutrient discharge. *Transactions of the ASAE* 29(4): 1037–1041.

PART THREE

Control Measures for Wind Erosion

3.1
Effects of Natural Savannah Windbreaks and Soil Ridging on Wind Erosion and Growth of Cowpea and Millet

Barbara Buerkert, Joachim Banzhaf, Andreas Buerkert, and
Dietrich E. Leihner

Introduction

Wind erosion is a major problem in several regions of the world, however, effective means have been developed to reduce wind erosion (Kort 1988; Tibke 1988). In semi-arid West Africa, wind speed and especially wind-blown soil particles may cause severe damage to young crops. The damage is due to partial or complete soil coverage of seedlings and abrasion to leaves. This can reduce plant stand and make replanting necessary (Chapter 2.2). Research on wind erosion control in the West African Sahel is rare. Wind barriers, consisting of trees and *Andropogon* grass, planted in rows perpendicular to the main storm direction effectively reduced the amount of wind-blown soil (Michels 1994; Chapter 3.2). However, the establishment of trees for wind erosion control is laborious and costly, therefore integrating agroforestry systems is difficult for subsistence farmers in Niger. Lamers and von Oppen (Chapter 4.1) calculated that farmers will need to invest 20 700 FCFA ha^{-1} in windbreaks in the first four years for seedlings, material, hired labor, and maintenance. Furthermore, it takes several years before planted tree windbreaks are effective in controlling wind erosion. In contrast, annual and perennial herbaceous wind barriers, used in the United States, were found to be a more flexible wind erosion management system than perennial woody windbreaks (Bilbro and Fryrear 1988). Farmers in Niger already know that barriers such as natural vegetation capture wind-blown soil particles (Chapter 4.3), however there is no scientific information about the effects of such windbreaks on crop growth and soil erosion in West Africa.

B. Buerkert, B. E. Allison and M. von Oppen (eds.), Wind Erosion in Niger, 87–104.

The extent of wind erosion depends not only on wind speed, but also on the susceptibility of a given soil. The predominantly sandy soils of the Sahel, derived from sand dunes (Chapter 1.2), appear to be highly susceptible to wind forces. Mulch applications of crop residues were effective in protecting sandy soils from erosion and in trapping wind-blown material (Chapters 3.3 and 3.4). However in Nigerien farming systems, the amount of crop residues available is often limited and their use for soil conservation purposes competes with other needs, such as building material, fodder, and firewood (Chapter 4.3).

Another effective means of reducing wind erosion on susceptible soils is soil tillage. Roughening the soil surface by ridging lowers the wind speed near the soil surface, thus, the most erodible soil fractions are protected (Armbrust *et al.* 1964). Field observations and wind tunnel experiments in the USA revealed that ridges of 0.13 to 0.25 m height may reduce the soil loss on sandy soils by up to 90% (Fryrear 1984). Under conditions in the southern Sahel, ridging is an effective way to increase pearl millet (*Pennisetum glaucum* (L.) R. Br.) and cowpea (*Vigna unguiculata* (L.) Walp.) establishment and grain yield (ICRISAT 1986; Klaij and Hoogmoed 1993). There is little scientific information on the effectiveness of soil ridging for wind erosion control.

A field experiment was carried out from 1985 to 1987 in southwestern Niger to investigate the effects of soil ridging and natural savannah windbreaks on wind speed, wind-blown soil, soil water content, and the growth of millet and cowpea. This chapter summarizes the results published previously by Banzhaf *et al.* (1992) and Leihner *et al.* (1993).

Experimental Setup

The study was conducted on a 12 ha area of natural bush savannah at the ICRISAT Sahelian Center (Niger). Rainfall was above the long-term average of 545 mm in 1985 and 1986, but very low (363 mm) in 1987 due to a 4-week dry spell in June. Also, the number of sandstorms without subsequent rainfall was higher in 1987 than in 1985 and 1986. The experiment was arranged as a split-plot design with three blocks. The main treatment consisted of soil ridging with ridges 0.25 m high and spaced 0.75 m apart. Flat and ridged blocks were arranged side by side. The control was left untilled (flat control). Ridges were constructed by using an oxen drawn soil ridger (locally manufactured). Within the main plots, four windbreak spacings of 6, 20, 40, and 90 m were used and millet and cowpea were grown in the space between the windbreaks. The field layout for one main plot is shown in Fig. 3.1–1. The windbreaks were oriented perpendicular to the east, the main storm direction. The windbreaks consisted of naturally growing savannah vegetation (Plate 4) which was left in place in 5 m wide strips. The natural vegetation was composed of annual species such as *Cenchrus biflorus*, *Cassia mimosoides*, *Stylosanthes* spp., and *Tephrosia* spp.. The perennial species were dominated

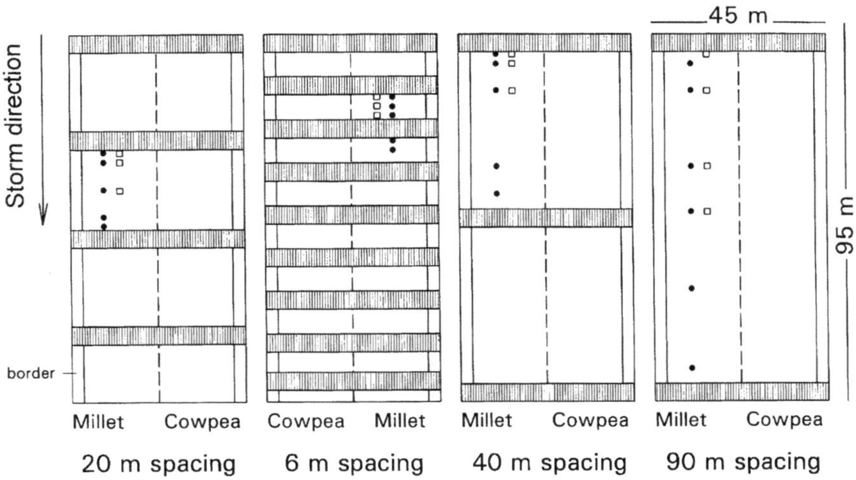

Fig. 3.1–1. Schematic diagram of field layout. Symbols indicate locations where measurements of wind speed (•) and subsoil moisture (□) were taken in both flat and ridged plots. (*Source*: Banzhaf *et al.* 1992, modified.)

by *Andropogon gayanus* Kunth. and *Guiera senegalensis.* The annual species created a rather homogeneous wind barrier of 0.6 m height (*H*), thus, in this study 1 *H* equaled 0.6 m. The above-ground dry matter yield at flowering was 2 t ha^{-1} and indicated a high production from the annual windbreak vegetation. *Guiera* bushes and *Andropogon* grass were randomly dispersed within the windbreaks and reached a height of 2.5 to 3 m.

Additional afforestation was done within the windbreak strips in 1985 and 1986 to accelerate the natural succession. Eight-month-old seedlings of local native trees such as *Faidherbia albida* (Del.), *Acacia nilotica* var. adansonii, and *Bauhinia rufescens* Lam., were transplanted in early to mid-July 1985. The seedlings were not watered, fertilized, or treated with pesticides in order to mimic the labor and financial constraints of farmers.

Cup anemometers were installed in 1987 at 0.3 m height, as indicated in Fig. 3.1–1, to measure wind run and wind speed. Five days before sowing, hourly readings were taken in millet plots from 0700 to 1700 h. The control wind speed was determined at 0.7 m height above the ground in the most eastern windbreak plots. Average wind speed from 10 days before sowing to 20 days after sowing was determined by installing anemometers in the center of the millet plots and recording total wind run.

Wind-blown soil was collected at 0.05 to 0.5 m height above the ground with modified Bagnold Sandcatchers (De Ploey and Gabriels 1980) for 45 days between early June and late July 1987. Soil water content was monitored at depths of 0 to 0.25 m gravimetrically and at depths of 0.35, 0.70, 1.80, and 2.40 m with a neutron probe at different locations within the millet plots (Fig. 3.1–1).

Millet, *cv.* CIVT, was sown with a hoe by placing 20 to 40 seeds in each sowing pocket and achieving a pocket density of 10 000 pockets ha^{-1}. Seedlings were thinned to three plants per pocket at 25 days after sowing (DAS). Cowpea was sown after millet by placing 5 to 10 grains in each sowing pocket. In 1985, 30 kg ha^{-1} of nitrogen (N) as calcium ammonium nitrate was applied before sowing. In 1986 and 1987, 15 kg N ha^{-1} and 13 kg ha^{-1} of phosphorus as triple superphosphate were applied before sowing and 30 kg N ha^{-1} at tillering. In 1987, 25 kg ha^{-1} of potassium were also applied as potassium chloride.

The above-ground dry matter yield was determined ten days after the onset of millet flowering and at physiological maturity by sampling ten pockets of millet and 20 pockets of cowpea. Additionally, in 1987 millet plants were harvested at thinning (5-leaf stage; 25 DAS). Early ground cover of cowpea was estimated at 40 and 50 DAS by placing a grid on the soil surface (Leihner *et al.* 1993). The leaf area of harvested millet samples was determined with a leaf area meter and the leaf area indices (LAI) were calculated. All yield data are based on the crop area harvested, because the land used by the windbreaks also produced other marketable goods such as fodder, firewood, and building material. Procedures for statistical analysis are given by Banzhaf *et al.* (1992) and Leihner *et al.* (1993).

Soil Ridging

The wind speed and the amount of wind-blown soil in ridged plots were not significantly different from the control (flat plots) at the beginning of the rainy season (data not shown). However, as the windbreak spacing decreased from 90 to 6 m, the amount of wind-blown soil particles in ridged plots was reduced by up to 50% (Fig. 3.1–2). There was a larger reduction in wind-blown soil in the flat treatments. In wind tunnel experiments, Armbrust *et al.* (1964) reported that ridges of dune sand from 51 to 203 mm high were effective in reducing wind speed, compared with a flat surface. Ridging with a fine sandy loamy soil reduced soil loss by about 90% with ridges from 63 to 254 mm high (Fryrear 1984). The effectiveness of the ridges in reducing wind erosion varies with their height, the degree of soil cloddiness, and the drag velocity (Armbrust *et al.* 1964). The smaller effects of soil ridges (compared with flat plots) on wind erosion control under Sahelian conditions, reported in this study, may have partly been caused by the low cloddiness of the sandy soil, ridge height, and wind flow patterns through the heterogeneous windbreaks.

The dry matter production of millet at thinning and ground cover of cowpea at 40 and 50 DAS were not affected by ridging (data not shown). Also at flowering, the overall effect of ridging on millet dry matter production and

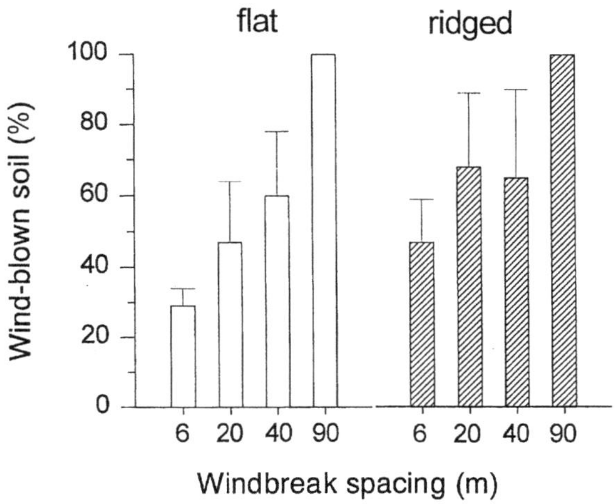

Fig. 3.1–2. Effect of windbreak spacing on the relative amount of wind-blown soil (at 0.05 to 0.5 m height above ground) in the center of flat and ridged treatments at the ICRISAT Sahelian Center, Niger in 1987. The 90 m windbreak spacing is considered the control (= 100%). Bars indicate one standard deviation. (*Source*: Banzhaf *et al.* 1992, modified.)

leaf area index was not significant (Table 3.1–1). In 1986, however, the stem and the total dry matter yield increased in ridged plots by 113 and 81%, respectively, compared with flat plots. The leaf dry matter yield and the leaf area index also tended to be higher. At maturity, differences in total dry matter yield of millet were smaller (24%) and not significant. In 1985 and 1987, dry matter yields at maturity were not significantly increased by ridging.

In the cowpea plots, soil ridging significantly increased total dry matter yield at flowering by 66% in 1985 and by 70% in 1986, compared with the flat control (Table 3.1–2). Although these effects due to ridging were smaller at maturity in 1986, there were 26% more pods per pocket and grain yield increased by 45% on ridged plots compared with flat plots. In contrast, the number of pods per pocket and grain yield tended to be smaller with ridging than with the flat treatment in 1987, the year with below normal rainfall.

Recently, it was shown that soil ridging increased millet seedling survival by 18% and increased millet grain yields significantly (53%) compared with the untilled control (Klaij and Hoogmoed 1993). Also, grain yields in cowpea were 10% and straw yields were 17% higher on ridged plots, and water use efficiency increased by 9.5% compared with untilled plots (ICRISAT 1986). The effects of soil ridging on crop production in the southern Sahel are generally attributed to an improvement in the physical characteristics of the soil. Soil ridging tends to break soil crusts, thereby improving infiltration rates and increasing the water available to plants. On the other hand, the destruction of ridges, by strong sandstorms followed by rains with high rainfall intensity

Table 3.1–1. Effects of soil ridging on millet dry matter yield (in kg ha^{-1}) at flowering and maturity, and on leaf area index (LAI) from 1985 to 1987

	FLOWERING			LAI	MATURITY		
	Dry matter				Dry matter		
Ridging	Stem	Leaf	Total		Straw	Grain	Total
1985							
flat	1910	660	2570	1.6	3060	790	3850
ridged	1770	590	2360	1.5	2750	680	3430
1986							
flat	620	610	1240	1.0	2320	530	2840
ridged	1320	940	2240	1.6	2870	660	3530
1987							
flat	570	400	970	0.6	1070	220	1290
ridged	480	340	820	0.5	970	210	1180
Analysis of variance							
Year	***	***	***	***	***	***	***
Ridging	n.s.	n.s.	n.s.	n.s.	n.s.	n.s.	n.s.
$Y^a \times R^b$	*	n.s.	*	n.s.	n.s.	n.s.	n.s.

(*Source*: Leihner *et al.* 1993)
*** and * F test significant at the 0.001 and 0.05 probability level, respectively
n.s. = not significant (= P > 0.1)
[a] Year
[b] Ridging

(Plate 13), and the delay in sowing due to ridging may be responsible for the absence of ridging effects in some years. Since rainfall at the beginning of the rainy season is often very limited and the distribution is variable (Chapter 1.1), sowing at the right time is crucial in the Sahelian environment. A delay in sowing after the first rain due to ridging, may offset the potential yield increases due to ridging in six out of ten years (Klaij and Hoogmoed 1993). However, in years when sandstorms cause the partial or complete coverage of young crop plants with soil, ridging may improve seedling establishment, and thus, crop yields.

Table 3.1–2. Effects of soil ridging on cowpea dry matter yield at flowering, pods per pocket, and grain yield at maturity from 1985 to 1987

	FLOWERING			MATURITY	
	Dry matter (kg ha^{-1})			Pods per	Grain
Ridging	Stem	Leaf	Total	pocket	yield (kg ha^{-1})
1985	*	*	*	n.s.	n.s.
flat	320	220	530	38	750
ridged	550	340	880	49	860
1986	*	n.s.	+	+	+
flat	630	560	1190	39	730
ridged	1250	770	2020	49	1060
1987	n.s.	+	n.s.	n.s.	n.s.
flat	200	140	340	7	90
ridged	220	160	380	5	70
Analysis of variance					
Year	***	***	***	***	***
Ridging	**	+	*	n.s.	*
$Y^a \times R^b$	n.s.	n.s.	n.s.	n.s.	n.s.

(*Source*: Leihner *et al.* 1993)
***, **, *, + F test significant at the 0.001, 0.01, 0.05, and 0.1 probability levels, respectively
n.s. = not significant (= P > 0.1)
[a] Year
[b] Ridging

Natural Savannah Windbreaks

Afforestation

The establishment rate of tree seedlings was low. Two years after planting, 32% of the *F. albida* seedlings, 36% of the *A. nilotica* seedlings, and 40% of *B. rufescens* seedlings had survived. Since rainfall was plentiful in July and August of 1985 and 1986, the low survival rate of seedlings was likely due to the lack of external inputs, such as fertilizers and pesticides, rather than to water shortages. Also, the 8-month-old seedlings may have been too old for transplanting. In contrast, survival rates of transplanted 3-month-old tree seedlings were almost 100% when water, fertilizer, and pesticides were added (Lamers *et al.* 1994). Under on-farm conditions however, the protection

Table 3.1–3. Shoot length (in m) of *Faidherbia albida*, *Acacia nilotica* and *Bauhinia rufescens* seedlings planted within the savannah windbreak strips, in 1986 (25 July) and 1987 (23 November) (means ± standard deviation; n = 100)

Tree species	Shoot length (m)	
	1986	1987
Faidherbia albida	0.29 ± 0.08	0.30 ± 0.13
Acacia nilotica	0.40 ± 0.11	0.86 ± 0.38
Bauhinia rufescens	0.72 ± 0.17	1.44 ± 0.66

of seedlings from browsing livestock and weeds must also be considered in order to guarantee a good tree establishment. In an on-farm study, 55% of the planted *A. gayanus* and 70% of the *Ziziphus mauritania* seedlings were lost due to browsing by livestock (ICRISAT 1992). Nigerien farmers also stated that *F. albida* needs to be protected against competition by weeds, particularly during early growth (Chapter 4.3).

Large differences were measured in early shoot growth among the tree species. While *B. rufescens* and *A. nilotica* doubled their shoot length within one year, the growth of *F. albida* trees was slow (Table 3.1–3). Similar growth differences among local tree species have also been reported (Lamers *et al.* 1994). Four years after planting, total annual dry matter production of *B. rufescens* increased threefold and *A. nilotica* twofold compared with the annual dry matter production of *F. albida*. These findings imply that growth characteristics of different tree species need to be considered when planning windbreak projects.

Wind Speed and Wind-blown Soil

The wind speed at 3 m distance leeward from the windbreaks was reduced by up to 70% of the control wind speed (Fig. 3.1–3). At 10 m leeward distance from the windbreaks, relative wind speed was reduced to 80% of the control, which was taken as the threshold for effective wind erosion control in this study. This means that the reduction in wind speed was effective to 17 times the windbreak height (17 *H*). Average relative wind speed in the center of the windbreak plots was reduced by up to 25% as the distance between the windbreaks decreased from 90 to 6 m (Fig. 3.1–4). Tibke (1988) noted that the effectiveness of a wind speed reduction increased exponentially. Thus, a small reduction in wind speed, as shown for windbreaks spaced 6, 20, and 40 m, may cause a substantial decrease in wind erosion in plots with windbreaks (Skidmore and Hagen 1977). While average wind speed above the windbreaks

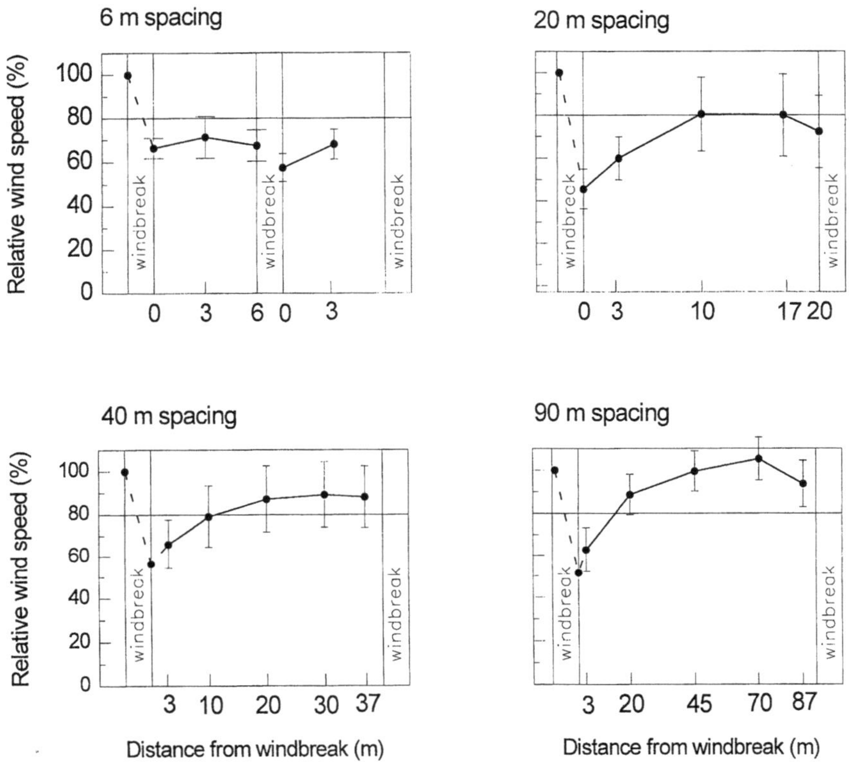

Fig. 3.1–3. Effect of windbreak spacing on relative wind speed at the 0.3 m height above the flat bare soil surface. Wind speed at the 0.7 m height (0.1 m above windbreak vegetation) referred to as control (= 100%). Data points are daily averages. Bars indicate one standard deviation. (*Source*: Banzhaf *et al.* 1992, modified.)

was 3 m s^{-1} in our study, wind speeds can frequently exceed 6 m s^{-1} during convective storms (Michels 1994). It is known, however, that windbreaks are effective in controlling wind erosion at a broad range of wind speeds. Wind speeds from 1.9 to 9.5 m s^{-1} were reduced by 45% in wheatgrass (*Agropyron elongatum*) barriers within a distance of 18.4 *H* from the windbreak (Aase *et al.* 1985). Thus, it is assumed that the reduction in wind speed with savannah windbreaks may be similar for strong erosive winds.

The savannah windbreaks significantly reduced wind-blown soil particles by 70% (6 m spacing) and 53% (20 m spacing) in flat plots (Fig. 3.1–2). These data agree with results for *A. gayanus* and *B. rufescens* windbreaks spaced at 30 m (Chapter 3.2). The annual soil flux was significantly reduced at 5 and 10 m distance from the windbreaks compared with the unprotected control (Chapter 3.2). Windbreaks can also trap wind-blown soil particles. Within a 4-year period, the soil surface increased 0.15 m in height with 5 m spaced *A. gayanus* windbreaks compared with the unprotected soil surface

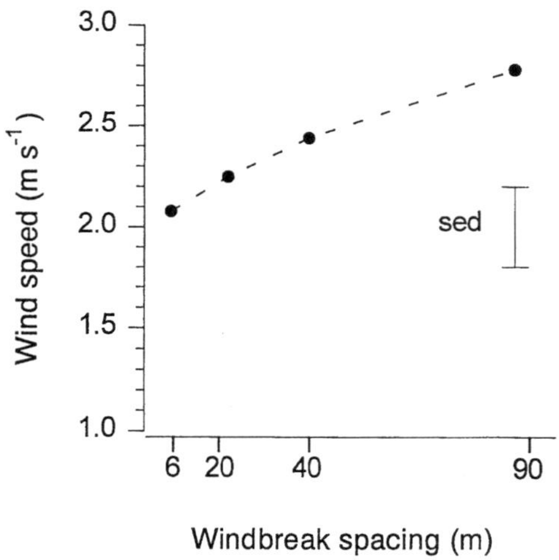

Fig. 3.1–4. Effect of windbreak spacing on wind speed. Data points are averages of measurements at the 0.3 m height in the center of flat and ridged treatments between 10 days before and 20 days after sowing; sed = standard error of the difference. (*Source*: Banzhaf *et al.* 1992, modified.)

(ICRISAT 1991). To fully quantify the effects of windbreaks on soil loss or deposition, thus soil fertility, topographic measurements need to be taken and the soil chemical status determined.

Soil Water Content

Independent of windbreak spacing, 5 to 17 mm of water were stored at 0 to 0.25 m soil depth (data not shown) in millet plots during the 1987 growing season. At soil depths of 0.35 to 2.40 m, the water content at the beginning of the rainy season (23 June) was significantly lower in plots with windbreaks spaced 6 m apart than in other windbreak treatments (Fig. 3.1–5). During the rainy season, windbreak spacing had no significant effect on the subsoil water content. In the early dry season (23 November), the subsoil water content in windbreak plots spaced 6 m apart was significantly lower than in plots of the other windbreak spacings. Additionally, the subsoil water content at the windbreak-crop interface was always lower compared with the center of the millet plots for all treatments (Banzhaf 1988). In 1985 as well as 1986, both years with greater total rainfall and more even distribution, no significant differences in subsoil water content among the windbreak treatments were found.

It is assumed that the transpiration of annual windbreak species is low during the dry season. Thus, occasional bushes and trees within the natural windbreak vegetation as well as tree seedlings that were planted, may have

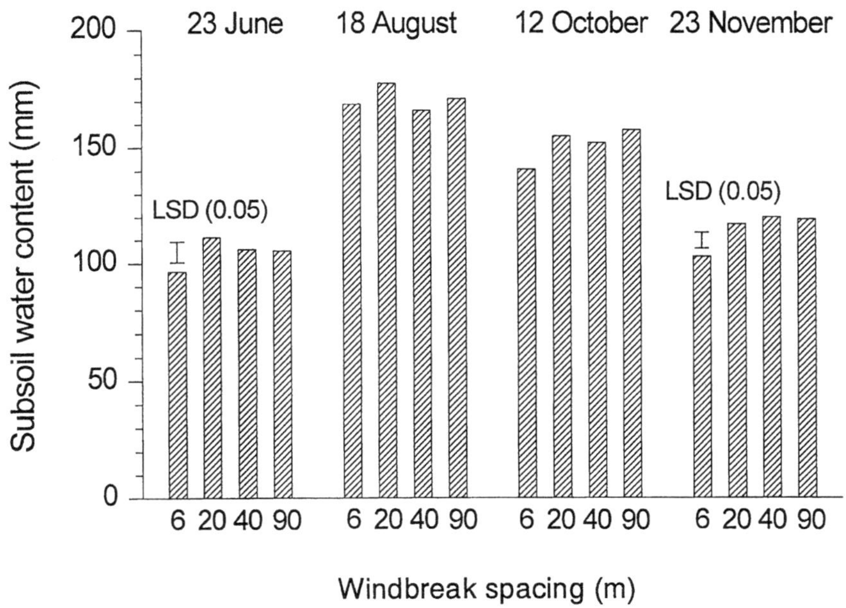

Fig. 3.1–5. Effect of windbreak spacing on subsoil water content. Columns are soil profile averages in the flat plot treatments and at different distances from the windbreak; LSD 0.05 = least significant difference at $P > 0.05$. (*Source*: Banzhaf *et al.* 1992, modified.)

been responsible for the lower water content in the soil during this period. Since water resources at the beginning of the rainy season are an important factor for early crop growth, it is likely that competition for soil water affected the growth of cowpea and millet, particularly in plots with the 6 m windbreak spacing. With respect to soil moisture, an ideal windbreak vegetation should have a high water-use efficiency (Smith *et al.* 1996) and a taproot system in order to avoid competition for water with nearby crops. The data presented suggest that there is competition for soil water at the windbreak-crop interface, thus, the overall impact on millet and cowpea growth in the Sahel will depend primarily on rainfall conditions, as well as windbreak species, height, and spacing.

Effects of Windbreaks on Millet and Cowpea Growth and Yield

Early Crop Growth

Millet dry matter yield at the 5-leaf stage (Table 3.1–4) and percentage of cowpea ground cover at 40 and 50 DAS (Table 3.1–5) were highest in plots with windbreaks spaced 6 m apart and decreased gradually with wider windbreak

Table 3.1–4. Effects of savannah windbreak spacing on millet dry matter yield (in kg ha^{-1}) at the 5-leaf stage (25 DAS), flowering, and maturity; and leaf area index (LAI) at flowering from 1985 to 1987

	25 DAS	FLOWERING				MATURITY		
	Dry matter	Dry matter			LAI	Dry matter		
Windbreak		Stem	Leaf	Total		Straw	Grain	Total
1985	n.d.	+	+	+	n.s.	n.s.	n.s.	n.s.
6 m	n.d.	1950ab	610ab	2560ab	1.5	2800	730	3530
20 m	n.d.	2500a	750a	3250a	1.7	2940	720	3660
40 m	n.d.	1570ab	620ab	2180ab	1.5	2910	700	3610
90 m	n.d.	1350b	520b	1870b	1.3	2970	800	3760
1986	n.d.	n.s.	n.s.	n.s.	n.s.	n.s.	n.s.	n.s.
6 m	n.d.	950	730	1680	1.2	2440	520	2960
20 m	n.d.	930	780	1700	1.3	2820	670	3490
40 m	n.d.	1000	810	1810	1.4	2470	570	3040
90 m	n.d.	1010	780	1790	1.3	2660	610	3270
1987	*	n.s.	n.s.	n.s.	n.s.	n.s.	n.s.	n.s.
6 m	33.5b	640	410	1060	0.6	970	140	1110
20 m	26.1ab	440	350	800	0.5	1020	230	1240
40 m	20.1a	570	380	950	0.5	1160	270	1430
90 m	17.6a	450	320	770	0.4	930	220	1150
Analysis of variance								
Year	n.d.	***	***	***	***	***	***	***
Windbreak	n.d.	n.s.	n.s.	n.s.	n.s.	n.s.	n.s.	n.s.
$Y^a \times WB^b$	n.d.	*	n.s.	+	n.s.	n.s.	n.s.	n.s.

(*Source:* Leihner *et al.* 1993)
***, *,+ F test significant at the 0.001, 0.05, and 0.1 probability level, respectively
n.s. = not significant (= $P > 0.1$), n.d. = not determined
Within columns, means followed by the same letter do not differ significantly using a Tukey test ($P < 0.05$ and 0.1, respectively)
[a] Year
[b] Windbreak spacing

spacings. Millet is very sensitive to wind erosion at early stages of development and sandstorms in the Sahel can result in the coverage of millet plants with wind-blown soil. Severe erosion conditions might necessitate replanting the millet field (Chapter 2.2). Although coverage of pockets was not quantified in this study, the reduced growth of young millet and the reduced ground cover of cowpea (Tables 3.1–4 and 3.1–5) with wider windbreaks spacings may be attributed to damage caused by sandstorms. The effect of windbreak treatments on the early crop growth supports results reported by Brenner

Table 3.1–5. Effects of windbreak spacing on cowpea ground cover at 40 and 50 days after sowing (DAS), dry matter yield at flowering, and grain yield at maturity from 1985 to 1987

	DAS		FLOWERING			MATURITY
	40	50	Dry matter (kg ha^{-1})			Grain yield
Windbreak	Ground cover (%)		Stem	Leaf	Total	(kg ha^{-1})
1985			n.s.	+	n.s.	n.s.
6 m	n.d.	n.d.	500	310a	810	700
20 m	n.d.	n.d.	460	280ab	720	790
40 m	n.d.	n.d.	350	250b	590	930
90 m	n.d.	n.d.	420	270ab	690	800
1986			n.s.	n.s.	n.s.	n.s.
6 m	n.d.	n.d.	930	690	1620	850
20 m	n.d.	n.d.	870	700	1570	1040
40 m	n.d.	n.d.	1010	650	1650	880
90 m	n.d.	n.d.	950	630	1580	810
1987			**	+	*	n.s.
6 m	37b	52b	330a	220a	550a	60
20 m	32ab	39a	192b	145ab	337b	70
40 m	22a	28a	150b	122b	272b	100
90 m	21a	28a	150b	120b	270b	90
Analysis of variance						
Year	n.d.	n.d.	***	***	***	***
Windbreak	n.d.	n.d.	n.s.	n.s.	n.s.	n.s.
Y[a] × WB[b]	n.d.	n.d.	n.s.	n.s.	n.s.	n.s.

(*Source*: Leihner *et al.* 1993)
***, **, *, + F test significant at the 0.001, 0.01, 0.05, and 0.1 probability level, respectively
n.s. = not significant (= $P > 0.1$), n.d. = not determined
Within columns, means followed by the same letter do not differ significantly using a Tukey test ($P < 0.1$)
[a] Year
[b] Windbreak spacing

(1991). He found that after sandstorms, viable leaf area and dry matter of young millet plants were increased in plots with neem (*Azadirachta indica*) windbreaks compared with plants in the unprotected control plots.

Wind tunnel studies showed that the survival of cowpea seedlings was not affected by wind and wind-blown soil. Plant fresh weight and grain yields were actually increased after sandblasting (Downes *et al.* 1977). Under

Sahelian conditions, the decrease in percentage of cowpea ground cover with wider spaced windbreaks may point to a slower regrowth of the young plants after damaging erosion events.

Above-ground Dry Matter Production

Averaged over the three years, the dry matter yields of millet and the leaf area indices were unaffected by windbreak spacing (Table 3.1–4). However, effects were observed in individual years. When compared with the 90 m windbreak spacing, the total dry matter yield of millet at flowering increased by 37% in plots with 6 m spacing and by 74% in plots with the 20 m windbreak spacing in 1985. The leaf area index was generally lowest in plots with the 90 m windbreak spacing. Similarly, in 1987 millet dry matter yield at flowering was 38% larger in plots with the 6 m windbreak spacing. However, these effects were not sustained at millet maturity. Such effects of windbreak spacing on early dry matter production were also observed for cowpea (Table 3.1–5). Brenner (1991), in his research with neem windbreaks, also reported that at the final harvest, total dry matter of millet was not significantly different from the control (no windbreaks). Therefore, one explanation may be that plants were able to compensate for early wind erosion damage by having higher growth rates later in the growing season. It was also observed in 1987 that millet plants, especially those from the narrow spaced windbreak plots, matured earlier than the local millet cultivars in surrounding fields. Thus, birds migrated to the ICRISAT experimental site and heavily attacked millet heads, thereby causing yield reductions.

Moreover, it appears that other constraints of crop production such as short-term water or nutrients deficits, especially nitrogen and phosphorus, may have offset the treatment effects at later growth stages, particularly in plots with windbreaks spaced 6 m apart. Brenner (1991) reported that the competition for water between a 6-year-old neem (*A. indica*) windbreak and an adjacent millet crop was evident up to a distance of 1.5 H from the windbreak. In the natural savannah windbreak plots, the competition for water between the windbreaks and the adjacent millet crops may have only been important in the dry year 1987, where there was less subsoil water at the windbreak-crop interface than in the center of millet plots. Since nutrients are one of the most limiting factors in the Sahel for the production of both natural savannah vegetation (Penning de Vries and Djitèye 1982) and crops (Pieri 1989; Bationo *et al.* 1990; Payne *et al.* 1991), it is likely that during the three years of the experiment the competition for nutrients adversely affected dry matter development in the proximity of windbreaks at later growth stages, and may have caused the variable millet and cowpea growth.

However, when evaluating the overall benefits of windbreaks in the Sahel, the use of numerous by-products derived from windbreaks such as construction materials, fodder, and firewood should be considered. In the Majjia

Valley, where a high water table provides excellent growing conditions for windbreak trees, the production of marketable wood derived from two-row neem (*A. indica*) windbreaks yielded about 880 US$ km^{-1} of windbreak for the first cut. In subsequent years, 270 US$ km^{-1} of windbreak could be obtained annually (Long *et al.* 1986).

While such planted windbreaks may provide marketable products, they require high investments for their establishment and maintenance (Chapter 4.1). The savannah windbreaks, in contrast, may be attractive to local farmers due to their easy establishment and their low cost maintenance, even if they require large areas of productive land. In order to decrease competition with productive land, savannah windbreaks could also be established on unproductive land such as field borders.

Summary and Conclusions

Soil ridging may not be effective in reducing soil erosion and increasing millet yields on the sandy soils of the southern Sahel. However, in some years it may improve cowpea growth and increase cowpea yields. In other years, labor and time constraints required to construct the ridges at planting time may outweigh potential increases in yield. The prerequisite for a widespread adoption of ridging would be the use of animal traction which is, at present, very uncommon in Niger and difficult to accomplish in the traditional farming systems. Among the reasons for limited adoption of animal traction by subsistence farmers in Niger, are the lack of financial resources and appropriate equipment (Chapter 1.3), but also the weak physical status of draft animals at the end of the dry season. These constraints would have to be removed to make ridging attractive to farmers in Niger and other Sahelian countries.

Natural savannah windbreaks are easy and cheap to establish and maintain. Spaced at 6 and 20 m, they may slow the decrease in soil productivity caused by wind erosion, and in the long-run help prevent desertification. It appears that the potential reduction of the direct and short-term impact of wind erosion on crop productivity is outweighed by competition for water, nutrients and, as trees grow higher, light. Under the semi-arid conditions of the southern Sahel, natural savannah windbreaks are unlikely to lead to substantial yield increases, especially in millet. Since grain yield is a very important factor in determining the profitability of windbreaks, the establishment of natural savannah windbreaks on productive land appears to be, at least in a short-term perspective, uneconomical for subsistence farmers in southwestern Niger. However, in the view of long-term soil conservation for sustainable agriculture in the Sahel, there is a need to protect the sandy soils. As a first step, it may be more feasible for farmers to establish natural savannah windbreaks on unproductive sites or field borders, and to improve the traditional open parkland systems by fostering the natural regrowth of trees.

Abstract

In the Sahel, strong sandstorms at the onset of the rainy season can transport considerable amounts of wind-blown soil and cause severe damage to young crop plants. Soil ridging and natural savannah windbreaks were tested as measures to reduce the effects of wind erosion in a field study conducted from 1985 to 1987 at the ICRISAT Sahelian Center, Sadoré (Niger). The ridges were 0.25 m high and spaced 0.75 m apart and for the flat treatment the soil was left untilled. The windbreaks consisted of 5 m wide strips of natural savannah vegetation, spaced 6, 20, 40, and 90 m apart. Wind speed, wind-blown soil, soil water content, and dry matter yield of pearl millet (*Pennisetum glaucum* (L.) R. Br.) and cowpea (*Vigna unguiculata* (L.) Walp.) were measured throughout the growing seasons. As the spacing of natural savannah windbreaks narrowed from 90 to 6 m, average wind speed in the center of the millet plots was significantly reduced from 2.8 to 2.1 m s^{-1}. At the 6 m spacing, the amount of wind-blown soil decreased by 50% in the ridged and by 70% in the flat treatments. Average subsoil water contents at the beginning and at the end of the cropping season were less in the 6 m windbreak spacing than in the others. Windbreaks that were spaced 6 and 20 m apart had increases in millet dry matter yield of 90 and 48%, respectively, at the 5-leaf stage. However, these early effects were not sustained throughout the growing season, probably due to increasing competition for water and nutrients. The effects of soil ridging on wind speed and wind-blown soil, as well as on straw and grain yields of cowpea and millet, were small. Nevertheless, cowpea was more responsive to ridging than millet. Although low-cost savannah windbreaks did not significantly increase millet yields in the first three years of establishment, they may help stabilize long-term crop production by reducing soil erosion losses. An appropriate selection of the type of windbreak and additional planting of local tree species within the natural savannah windbreaks, may increase the effectiveness of the windbreaks in terms of wind erosion control and provide additional marketable commodities.

Résumé

Effets des brise-vent naturels de savane et du buttage du sol sur l'érosion éolienne et au croissance du niébé et du petit mil

Au début de la saison des pluies au Sahel, de fortes tempêtes de sable peuvent transporter des quantités considérables de sol causant ainsi de graves dommages aux cultures à leur premier stade de développement. Lors d'une étude de terrain menée de 1985 à 1987 au Centre Sahélien de l'ICRISAT, Sadoré (Niger), on a mesuré les effets du buttage du sol et des brise-vent de savane naturelle dans la réduction de l'érosion éolienne. Les sillons profonds de 0,25 m ont été espacés de 0,75 m sur terrain billoné, un autre terrain restant plat. Les brise-vent consistaient en des rangées larges de 5 m de végétation de savane naturelle, espacées de 6, 20, 40 et 90 m. La vitesse du vent, le sol transporté par le vent, les réserves en eau, et les productions de matières sèches ont été mesurés tout au long des saisons de croissance pour le petit mil (*Pennisetum glaucum* (L.) R. Br.) et pour le niébé (*Vigna unguiculata* (L.) Walp.). Lorsque l'espace entre les brise-vent a été ramené de 90 à 6 m, la vitesse moyenne du vent au centre des lopins de petit mil a été nettement réduite de 2,8 à 2,1 m s^{-1}. Avec un espacement de 6 m, la quantité de sol transporté par le vent a diminué de 50% en terrain billoné et de 70% sur terrain plat. Les réserves hydriques moyennes du sol en début et en fin de saison ont été inférieures avec les brise-vent espacés de 6 m par rapport aux autres brise-vent. Les brise-vent espacés de 6 et 20 m ont apporté une augmentation dans la production de 90 et 48% de matière sèche de petit mil précoce. Cependant, ces effets précoces n'ont pas duré tout au long de la saison de croissance, probablement du fait de la compétition croissante pour l'eau et les éléments nutritifs. Les effets du billonage du sol sur la vitesse du vent et sur le sol qu'il transporte, n'ont pas été significatifs. Par conséquent, les effets sur la production de matière sèche et le rendement en grains de niébé et de petit mil ne l'ont pas été non plus. Cependant, il apparait que la production de niébé répond mieux à la technique de buttage du sol que celle de petit mil. Bien que les brise-vent de savane bon marché n'aient pas augmenté de façon significative les rendements en petit mil pendant les trois premières années de son implantation, ils peuvent aider à stabiliser la production des cultures à long terme en conservant le sol. Une sélection appropriée du type de brise-vent et la plantation supplémentaire d'espèces locales d'arbres au sein des brise-vent naturels peuvent augmenter l'efficacité du contrôle de l'érosion éolienne et la production commercialisable.

References

Armbrust, D.V., W.S. Chepil, and F.H. Siddoway. 1964. Effects of ridges on erosion of soil by wind. *Soil Science Society of America Proceedings* 28: 557–560.

Aase J.K., F.H. Siddoway, and A.L. Black. 1985. Effectiveness of grass barriers for reducing wind erosiveness. *Journal of Soil and Water Conservation* 40: 354–357.

Banzhaf, J. 1988. Auswirkungen von Windschutzstreifen aus Brachevegetation auf Wachstum und Ertragsbildung von Perlhirse (*Pennisetum americanum* (L.) Leeke) und Cowpea (*Vigna unguiculata* (L.) Walp.) im südlichen Sahel Westafrikas. Diss. Universität Hohenheim, Stuttgart, Germany.

Banzhaf, J., D.E. Leihner, A. Buerkert, and P.G. Serafini. 1992. Soil tillage and windbreak effects on millet and cowpea. I. Wind speed, evaporation, and wind erosion. *Agronomy Journal* 84: 1056–1060.

Bationo, A., S.H. Chien, J. Henao, B.C. Christianson, and A.U. Mokwunye. 1990. Agronomic evaluation of two unacidulated and partially acidulated phosphate rocks indigenous to Niger. *Soil Science Society of America Journal* 54: 1772–1777.

Bilbro, J.D. and D.W. Fryrear. 1988. Annual herbaceous wind barriers for protection of crops and soils and managing snowfall. *Agriculture, Ecosystems and Environment* 22/23: 103–122.

Brenner, A.J. 1991. Tree-crop interactions within a Sahelian windbreak system. Ph.D. Diss. University of Edinburgh, U.K.

De Ploey, J. and D. Gabriels. 1980. Measuring soil loss and experimental studies. pp. 63–96. *In* M.J. Kirkby and R.P.C. Morgan (eds.) Soil erosion. John Wiley and Sons, Chichester, U.K.

Downes, J.D., D.W. Fryrear, R.L. Wilson, and C.M. Sabota. 1977. Influence of wind erosion on growing plants. *Transactions of the ASAE* 20: 885–889.

Fryrear, D.W. 1984. Soil ridges-clods and wind erosion. *Transactions of the ASAE* 27(2): 445–448.

Hoogmoed, W.B. and L. Stroosnijder. 1984. Crust formation on sandy soils in the Sahel. I. Rainfall and infiltration. *Soil and Tillage Research* 4: 5–23.

ICRISAT. 1986. ICRISAT West African Programme, Annual Report for 1985. International Crops Research Institute for the Semi-Arid Tropics, Niamey, Niger.

ICRISAT. 1991. ICRISAT West African Programme, Annual Report for 1990. International Crops Research Institute for the Semi-Arid Tropics, Niamey, Niger.

ICRISAT. 1992. ICRISAT West African Programme Annual Report for 1991. International Crops Research Institute for the Semi-Arid Tropics, Niamey, Niger.

Klaij, M.C. and W.B. Hoogmoed. 1993. Soil management for crop production in the West African Sahel. II. Emergence, establishment, and yield of pearl millet. *Soil and Tillage Research* 25: 301–315.

Kort, J. 1988. Benefits of windbreaks to field and forage crops. *Agriculture, Ecosystem and Environment* 22/23: 103–122.

Lamers, J.P.A., K. Michels, and R.J. Vandenbeldt. 1994. Trees and windbreaks in the Sahel: Establishment, growth, nutritive and calorific values. *Agroforestry Systems* 26(3): 171–184.
Leihner, D.E., A. Buerkert, J. Banzhaf, and P.G. Serafini. 1993. Soil tillage and windbreak effects on millet and cowpea. II. Dry matter and grain yield. *Agronomy Journal* 85: 400–405.
Long, S.P., N. Persaud, M. Ganhah, and M. Ouattara. 1986. Influence of a neem (*Azadirachta indica*) windbreak plantation on millet yields and microclimate in Niger, West Africa. pp. 183–184. *In* International Symposium on Windbreak Technology, Lincoln, Nebraska, 22–26 June 1986, Great Plains Agric. Counc. Publ. 117, Montana State Univ. Coop. Ext. Serv., Bozeman, MT.
Michels, K. 1994. Wind erosion in the southern Sahelian zone. Diss. Universität Hohenheim, Verlag Ulrich E. Grauer, Stuttgart, Germany.
Payne, W.A., R.J. Lascano, L.R. Hossner, C.W. Wendt, and A.B. Onken. 1991. Pearl millet growth as affected by phosphorus and water. *Agronomy Journal* 83: 942–948.
Penning de Vries, F.W.T. and M.A. Djitèye. 1982. La productivité des pâturages saheliens. Une étude des sols, des végétations et de cette ressource naturelle. Agric. Re. Rep. 918. Pudoc, Wageningen, The Netherlands.
Pieri, C. 1989. Fertilité des terres de savanes. Bilan de trente dans de recherche et de développement agricoles au sud du Sahara. Ministère de la Coopération et CIRAD-IRAT, La Documentation Française, Paris, France.
Skidmore, E.L. and L.J. Hagen. 1977. Reducing wind erosion with barriers. *Transactions of the ASAE* 20(5): 911–915.
Smith, D.M., P.G. Jarvis, and J.C.W. Odongo. 1996. Water-use by Sahelian windbreak trees. pp. 115–128. *In* B. Buerkert, B.E. Allison, and M. von Oppen (eds.) Wind erosion in West Africa: The problem and its control. Proceedings International Symposium, Hohenheim, Germany, 5–7 Dec. 1994, Margraf Verlag, Weikersheim, Germany.
Tibke, G. 1988. Basic principles of wind erosion control. *Agriculture, Ecosystems and Environment* 22/23: 103–122.

3.2
Effects of Planted Windbreaks on Wind Erosion and Millet Growth, and Biomass Production of the Windbreaks

Ludger Eltrop, Bruce E. Allison, John P.A. Lamers, and Karlheinz Michels

Introduction

The use of trees as a protection against wind erosion has a long tradition in West Africa (Le Houérou 1989; Sowers and Issoufou 1992) and worldwide (Rosenberg 1974). In the Sahel, farmers often take advantage of natural vegetation and use it for windbreaks. However, in the southern Sahel tree stocks continue to diminish due to an increasing demand for firewood and the use of marginal areas for food production. Currently, the Sahelian countries consume far more firewood than the total output of the accessible natural forests and parkland systems (Gorse and Steeds 1985). Additionally, in the low external input cropping systems, soil fertility and crop yields have decreased due to wind erosion damage and shorter or omitted fallow periods (Vierich and Stoop 1990). These factors contribute to the further degradation of the poor sandy soils in West Africa. Windbreaks are thought to alleviate these problems in the Sahel, therefore, their establishment was reinforced after the severe droughts in the early seventies (Guinaudeau 1987; Den Heijer 1990).

The principle aim of windbreaks is to reduce the speed of wind and, consequently, to decrease the damage to soils, plants, and animals (Bird *et al.* 1992). Windbreak trees and shrubs also provide useful material for households, construction, and firewood (Chapter 4.3). Significant amounts of wind-blown soil may be trapped in the windbreaks, thus decreasing soil degradation (Renard and Vandenbeldt 1990). However, windbreaks differ considerably in their effectiveness in reducing wind erosion depending on their height, density, shape, orientation, and length (Hagen and Skidmore 1971; Wilson and Cooke 1980). Accordingly, wind speed and associated parameters like air temperature, air humidity, and CO_2 concentration may vary significantly in the lee

B. Buerkert, B. E. Allison and M. von Oppen (eds.), Wind Erosion in Niger, 105–122.

of the windbreaks. The short- and long-term productivity and the ecological value of windbreak-cropping systems is thought to exceed that of a pure agronomic system (King 1989).

Results on the effect of windbreaks on crop yields are often contradictory (Ben Salem 1991; Kaisin 1994). Natural savannah windbreaks, spaced 6 and 20 m apart, decreased wind speed and the amount of wind-blown soil compared to windbreaks spaced 90 m apart (Chapter 3.1). The total dry matter yield of millet was increased by 48% in the 20 m windbreak plots and by 90% in the 6 m windbreak plots at thinning, but at the end of the growing period no significant difference in millet dry matter production was observed among the windbreak plots (Chapter 3.1). Brenner (1991) found increased millet grain yields in an on-station trial with 6 m high windbreaks of *Azadirachta indica*. However, other authors reported that crop production was not increased (Kessler and Breman 1991; Kaisin 1994), or that the increase was only due to unique site characteristics (Sumberg and Burke 1991). The main drawback for increasing crop yields in windbreak-cropping systems is assumed to be the competition of the windbreak plants with the crop species for light, water, and nutrients (Vandenbeldt 1989; Kerkhof 1990) or the allelopathic effects between incompatible partners (Yadav *et al.* 1993). For the establishment of new windbreak-cropping systems and their acceptance by local farmers, more detailed information on tree-crop interactions and on successful and economically valuable tree-crop combinations in the southern Sahel is needed.

The objective of the present study was to determine the effects of different windbreak species on wind erosion and the production of pearl millet in the adjacent fields. Special emphasis was given to the windbreak species *Andropogon gayanus* and *Bauhinia rufescens*. Furthermore, fodder and wood production of the windbreak species was measured to evaluate the economic potential of windbreak-millet cropping systems (Chapter 4.1).

Experimental Site and Cropping System

The field experiment was carried out during 1992 and 1993 at the ICRISAT Sahelian Center, Sadoré (Niger) (Michels 1994). Six tree species and one grass species were used: the caesalpinoid *Bauhinia rufescens* Lam. (Plate 14), the indigenous *Acacia nilotica* var. Adansonii, *Acacia senegal* (L.) Willd., the Australian *Acacia holosericea*, the neem tree *Azadirachta indica* A. Juss., the West African *Faidherbia albida* Del., and the perennial grass *Andropogon gayanus* Kunth (Plate 15). In 1988, the windbreaks were planted in 50 m long, double-staggered rows in a north-south direction, perpendicular to the predominant wind direction, and in three blocks, one behind the other. The distance between the centers of the windbreak rows was 30 m. Prior to the rainy season, the windbreaks were trimmed to a height of 2 m and a width of 3.5 m. Plots without windbreaks served as the control.

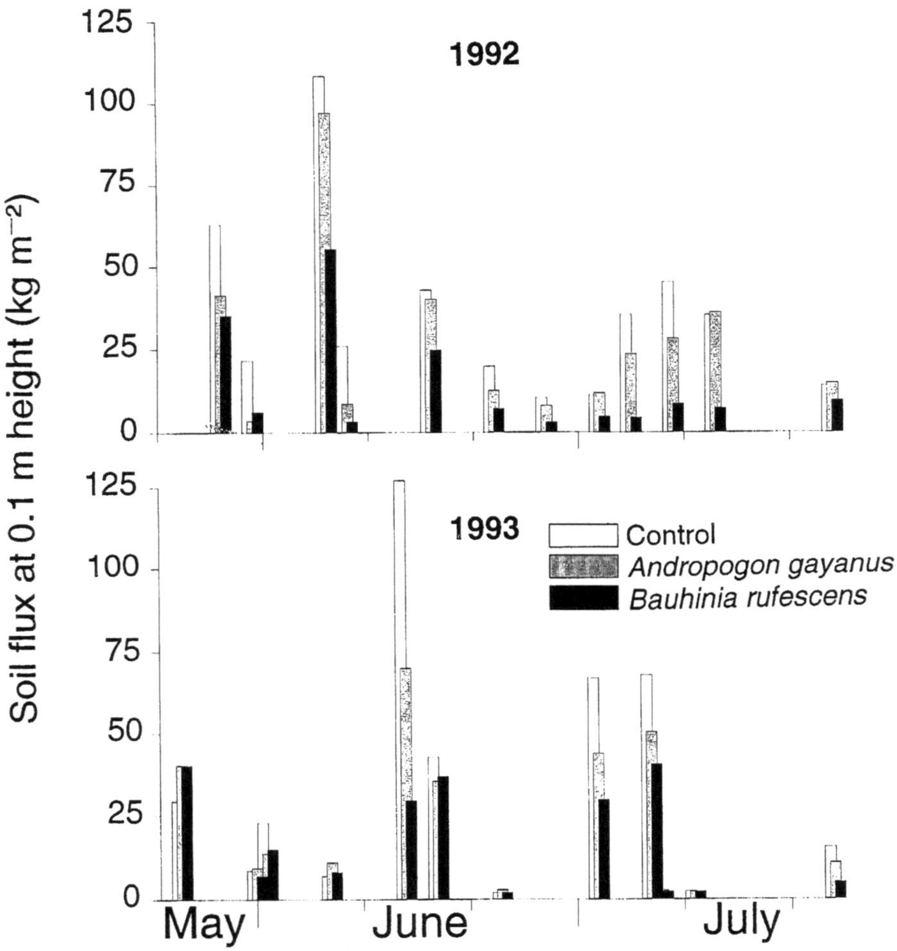

Fig. 3.2–1. Wind erosion events during the 1992 and 1993 growing seasons at the ICRISAT Sahelian Center (ISC), Sadoré (Niger). The soil flux is the amount of wind-blown soil captured during sandstorms in BSNE soil traps at 0.1 m height above ground level.

Pearl millet, *cv.* CIVT, was sown manually in planting holes ('pockets') spaced 1 by 1 m in each plot on 27 May (1992) and on 7 June (1993). The first millet row was sown at 2.75 m distance from the center of a windbreak. Three weeks after sowing, plants were thinned to three plants per pocket. Prior to the growing season, the millet field was fertilized with 19.5 kg ha^{-1} of phosphorus (P) as single superphosphate. At tillering, 30 kg nitrogen (N) ha^{-1} and at booting 15 kg N ha^{-1} were applied as calcium ammonium nitrate.

Table 3.2–1. Total soil flux at 0.1 m height above ground level in the growing seasons 1992 and 1993. Values are means of three sampling points at 5, 10, and 15 m distance from the edge of the windbreaks or on the control plots without windbreaks

Windbreak species	1992		1993	
	(kg m^{-2})	(%)	(kg m^{-2})	(%)
Control (without windbreaks)	433.6	100.0	394.5	100.0
Andropogon gayanus	326.4	75.1	292.3	73.9
Bauhinia rufescens	169.8	39.2	217.5	54.9

Windbreak Effects on Wind Erosion, Soil Flux, and Chemical Soil Properties

Wind Erosion and Soil Flux

BSNE soil samplers (Fryrear 1986) were used to collect soil at 0.1 m height above ground level (Plate 12). Twelve convective storms were counted during the rainy seasons, between May and the end of July, in 1992 and 1993. Similar to 1990 (Chapter 2.2), the storms occurred at irregular intervals and differed in duration and strength. However, the sandstorms in 1992 and 1993 were less frequent and less severe than those in 1990 (Chapter 2.2). In the majority of the erosion events, the soil flux was less in the windbreak plots than in the control plots without windbreaks (Fig. 3.2–1). The decrease was more pronounced with the tree species *B. rufescens* than with the grass *A. gayanus*. During the most severe sandstorm in 1992, 109 kg soil m^{-2} sampler opening were captured at 0.1 m above ground level. Compared to the control plots, the soil flux was reduced to 90% in the *A. gayanus* plots and to 51% in the *B. rufescens* plots. During the most severe sandstorm in 1993, the soil flux in the windbreak plots decreased to an even higher extent compared to the control plots.

The total amount of soil collected during the rainy season was significantly lower in the windbreak plots compared to the control plots (Table 3.2–1). The results demonstrate that soil flux during convective storms is effectively reduced by windbreaks and thereby plants may be protected from leaf damage. Since increasing wind speed exponentially increases wind erosion (Tibke 1988), a small reduction in wind speed on windbreak sheltered plots may significantly reduce wind erosion (Skidmore and Hagen 1977).

Even though the windbreaks were initially of similar height, the soil flux decreased less on the plots with *A. gayanus* than on those with *B. rufescens*. This can be attributed to differences in plant architecture (branching pattern, shape of leaves, etc.) and porosity of the two windbreak species. Throughout

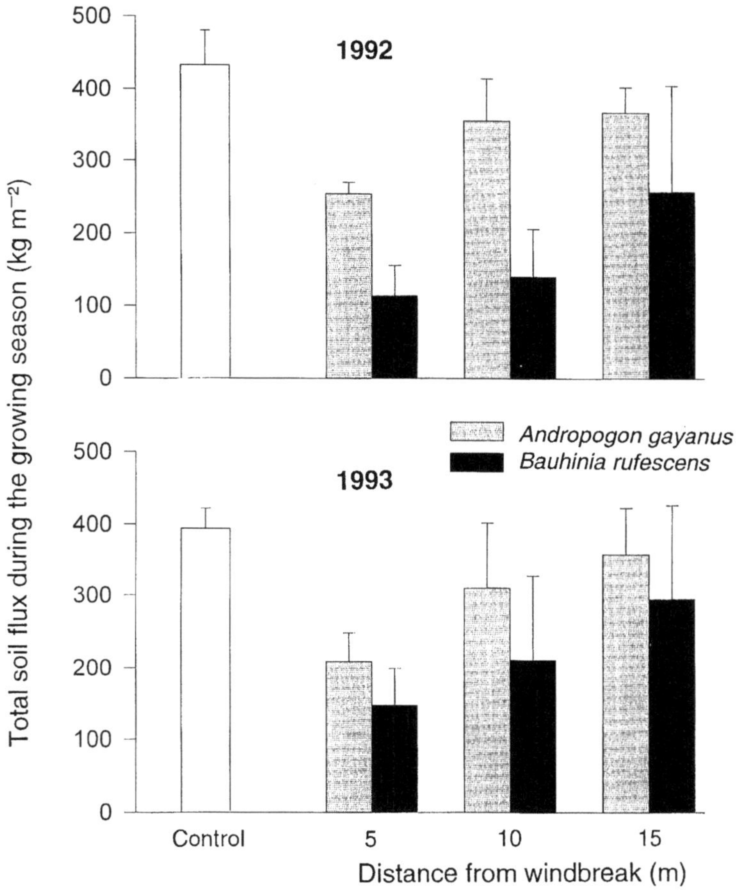

Fig. 3.2–2. Total soil flux at 0.1 m height above ground level in the growing seasons 1992 and 1993. Wind-blown soil was collected at different distances from the windbreak edge. Vertical bars indicate mean standard error (SE).

the years, the canopy density at 0.5 m height above ground level was higher for *B. rufescens* than for *A. gayanus* (Plates 14 and 15).

The effect of windbreaks on soil flux diminished with increasing distance from the windbreaks (Fig. 3.2–2). At 5 m distance, the total seasonal soil flux decreased significantly on the windbreak plots compared to the control. At 10 m distance, soil flux was significantly reduced only in the plots with *B. rufescens*, whereas at 15 m distance no significant reduction in soil flux was observed. The distance effectively protected by the windbreaks was thus five times the plant height (5*H*) with *A. gayanus* and seven times the plant height (7*H*) with *B. rufescens*.

Table 3.2–2. Chemical properties of soil samples from 0 to 0.1 m depth, collected in May 1993 at different distances (1, 3, and 10 m) from windbreaks with *Andropogon gayanus* and *Bauhinia rufescens* and from control plots without windbreaks

Parameter	Control	*Andropogon gayanus*			*Bauhinia rufescens*		
		Distance from windbreak			Distance from windbreak		
		1 m	3 m	10 m	1 m	3 m	10 m
pH_{KCl}	3.9	3.9	3.9	3.9	4.1	4.1	4.0
$C_{organic}$(%)	0.28	0.30	0.34	0.34	0.36	0.32	0.36
N_{total}(mg kg^{-1})	139.3	171.0	168.0	165.3	181.7	157.0	177.0
P_{Bray} (mg kg^{-1})	24.9	23.4	28.7	21.7	15.6	23.7	22.4
$K^{+}_{exch.}$ (cmol kg^{-1})	0.04	0.06	0.06	0.04	0.09	0.04	0.05
$Ca^{2+}_{exch.}$ (cmol kg^{-1})	0.34	0.38	0.42	0.41	0.48	0.41	0.47
$Mg^{2+}_{exch.}$ (cmol kg^{-1})	0.12	0.14	0.15	0.13	0.21	0.15	0.15
$Al^{3+}_{exch.}$ (cmol kg^{-1})	0.22	0.27	0.21	0.28	0.08	0.18	0.21
CEC ($cmol_c$ kg^{-1})	0.89	0.96	1.00	0.99	1.00	0.93	0.98

The extent of the area sheltered by windbreaks is determined by the shape and height of the barrier (Van Eimern *et al.* 1964; Rosenberg 1974; Raussen 1990). However, a precise relationship between the architecture of a windbreak and the sheltered area is difficult to establish. In the present study, a significant reduction in the amount of wind-blown soil was measured only up to 10 m distance from the windbreaks (Fig. 3.2–2). For annual windbreaks of low density, the protected area *H* was higher than for the tree species tested in the present study (Chapter 3.1). Rosenberg (1974) found that the protected area behind dense barriers extends from 10 to 15*H* downwind, whereas the protected area behind permeable windbreaks extends from 20 to 25*H*. Thus, a careful selection of suitable windbreak species and a well-planned management of the windbreak architecture are essential to increase the effectiveness of windbreaks in minimizing wind erosion.

Chemical Soil Properties

Five years after the installation of the windbreaks, the soil chemical properties in the 0 to 0.1 m soil horizon were only partly changed by the presence of windbreaks (Table 3.2–2). No significant differences in soil chemical properties attributable to the presence of windbreaks were observed in deeper soil horizons (data not shown). In the windbreak plots, particularly organic carbon (C), total N, and the cation exchange capacity (CEC) were significantly increased. The increase in organic C and N can likely be attributed to the input of organic matter from leaf litter. The concentration of exchangeable calcium (Ca) and magnesium (Mg) tended to be higher in the windbreak plots compared to the control, whereas the concentration of P and potassium (K) were not significantly different. Assuming that large quantities of dust, which contain high amounts of K and also considerable amounts of Ca and Mg (Chapter 2.1), may be trapped and deposited in the windbreaks, the topsoil concentrations of these elements, particularly in the vicinity of the windbreaks, are surprisingly low. However, the reasons for this are not clear.

The results presented are confirmed by those of Okorio (1992), who found significantly larger amounts of organic C, available P, and total N at 0 to 0.6 m soil depth with *F. albida* windbreaks in the semi-arid region of Morogoro (Tansania). Increased concentrations of organic N and exchangeable base cations under *F. albida* were also found in West Africa (Depommier *et al.* 1992).

The small increase in soil nutrient concentrations in the windbreak plots reported in this study may also be due to the mineral nutrient requirements of the windbreak species. For example, *B. rufescens* is known to extract considerable amounts of nutrients from the soil (Lamers *et al.* 1994b). Local farmers considered soil enrichment and following higher millet yields as one of the most important reasons for installing and maintaining windbreaks on their fields (Lamers *et al.* 1994a).

The higher concentrations of base cations and organic C may improve the long-term soil fertility status and also reduce aluminium (Al) toxicity. The Al concentration in the topsoil at 1 m distance from the *B. rufescens* windbreaks was distinctly lower than in the control plots (Table 3.2–2). The low Al concentration in this soil indicates a low potential for Al toxicity, despite the fairly low pH of 4.0. In general, higher contents of organic matter increase the complexation of toxic Al compounds (Hue *et al.* 1986) and the base cations stabilize or even increase the pH of the soil, thereby decreasing the solubility and concentration of potentially toxic Al. Besides increasing soil fertility directly, windbreaks also indirectly stimulate crop productivity by decreasing the concentration of potentially toxic Al.

In conclusion, in the windbreak plots, only small changes the soil chemical parameters were observed. Nevertheless, since the largest proportion of mineral elements are found at 0 to 0.1 m soil depth (Michels, 1994), it is particularly important to protect the topsoil from wind erosion.

Windbreak Effects on Growth of Pearl Millet

Establishment of Millet

Since the covering of millet pockets with wind-blown soil is one of the major damaging effects of sandstorms on millet production (Chapter 2.2), the number of pockets with viable millet plants *versus* sown pockets was counted to assess early plant establishment.

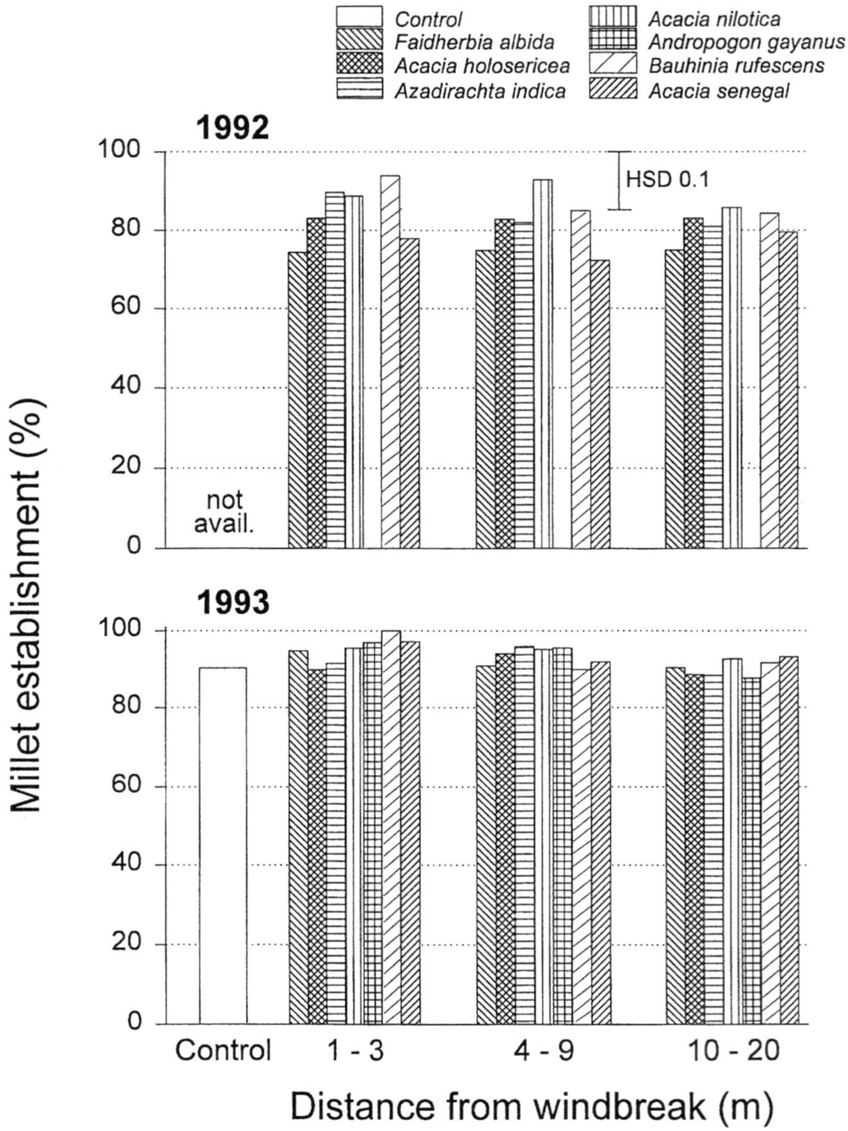

Fig. 3.2–3. The effect of windbreaks on the coverage of millet sown pockets with soil (number of viable millet versus the number of sown pockets) at three distances from the windbreaks in 1992 and 1993.

Although the amount of wind-blown soil clearly decreased in the windbreak plots (Table 3.2–1), there was no consistent difference between the control plots and the seven windbreak plots in the proportion of sown pockets covered with soil (Fig. 3.2–3). Additionally, the proportion of pocket cov-

Table 3.2–3. Straw and grain dry matter yields of pearl millet in 1992, grown in windbreak-cropping systems with seven different windbreak species. The control plots were without windbreaks. Values marked with * are significantly different from the control at $P > 0.05$

	Straw (kg ha^{-1})				Grain (kg ha^{-1})			
	Distance from windbreak (m)				Distance from windbreak (m)			
Windbreak species	1–3	4–9	10–20	Mean	1–3	4–9	10–20	Mean
Control	–	–	–	1219	–	–	–	683
Faidherbia albida	2255*	1894	1970*	2039*	1128*	986	1024*	1046*
Acacia holosericea	756	2035*	1831	1540	495	1157*	1043*	898
Azadirachta indica	712	1966*	1467	1785	320*	953	841	705
Acacia nilotica	880	1686	1321	1296	450	956	775	727
Andropogon gayanus	1250	1434	1047	1244	352	740	599	564
Bauhinia rufescens	723	1543	1430	1232	288*	882	684	618
Acacia senegal	468*	791	1305	855	254*	443	769	489

erage did not change with distance from the windbreaks and no difference was found between the years 1992 and 1993 (Fig. 3.2–3). This may be due to the late occurrence of the sandstorms after the young plants were already established. Most likely, the time of occurrence during millet establishment rather than the extent of the soil flux *per se* is the main factor determining the degree of crop damage (Chapter 2.2). It is also possible that due to turbulent windstreams between the windbreak rows, the wind-blown soil was equally re-distributed among the plots and that control plots were influenced by the adjacent windbreak plots. Since the conditions for crop establishment vary greatly from year to year in the Sahelian environment (Chapter 1.1), the protection of millet seedlings by windbreaks needs to be evaluated on a long-term basis. Windbreaks may be more effective in protecting young millet plants if more wind-blown soil is captured, for example, by carefully selecting tree species.

Dry Matter Yield of Pearl Millet

The dry matter yield of millet at physiological maturity was not consistently affected by the windbreaks in 1992 (Table 3.2–3). At a distance of 1 to 3 m from the windbreaks, the dry matter yield of millet straw and grain was highest in windbreak plots with *F. albida* and lowest in those with *A. senegal.* In general, dry matter production of both straw and grain tended to be lower in the vicinity of the windbreaks (between 1 and 3 m) compared to the other distances (Table 3.2–3), supporting similar results by Long and Persaud (1988) and Brenner (1991). In 1993, the differences in millet dry matter production among the treatments were similar to that of 1992, but due to heavy damage by the millet head caterpillar (*Heliocheilus* (= *Raghuva*) *albipunctella*) the grain yield was distinctly lower than in 1992 (data not shown). Other crop species gave different results. For example, with sorghum and maize dry matter yield increased when grown in soils taken from neem (*A. indica*) plantations, whereas the dry matter of groundnut remained unchanged (Verinumbe 1991). In wheat, grown on plots with *A. nilotica* as windbreaks, dry matter yield was not affected (Puri and Bangarwa 1992). This may be due to allelopathic reactions caused by components of leaf litter from *A. nilotica* (Yadav *et al.* 1993).

Windbreak-crop Competition

The increased millet yields on plots with *F. albida* and *A. holosericea* compensated for the loss of 11.7% of the experimental field occupied by the windbreaks. However, with the other windbreak species the decrease in grain yield must be added to the loss of arable land. Competition for light, nutrients, and water is considered an important factor for potential yield losses close to the windbreaks (Rosenberg 1974; Bird 1984). The water content of the soil close to neem (*A. indica*) windbreaks was lower than further away (Brenner 1991). During periods of low rainfall, the subsoil water contents were lower in narrow spaced windbreak plots compared to widely spaced plots (Chapter 3.1). Smith *et al.* (1996) stated that the water consumption of tree species in the Sahel may vary significantly and should be thoroughly considered before planting trees. In this study, the competition for limited water resources may have been an important factor in decreasing dry matter production of millet close to the windbreaks (Table 3.2–3). On the other hand, crop productivity in the southern Sahelian zone is more limited by low soil fertility, particularly by low P availability, rather than by water shortages (Bationo *et al.* 1990; Bley 1990; Payne *et al.* 1991). However, information about the competition between windbreaks and adjacent millet crops for nutrients and water is scarce and further research is needed.

The 'Faidherbia Albida Effect'

The dry matter yield of millet consistently increased on plots with *F. albida*. This effect, the so-called '*F. albida* effect' is well documented (Belsky *et al.* 1989; Vandenbeldt 1992; Vandenbeldt and Williams 1992). Due to the inverse life cycle, with the shedding of leaves at the beginning of the rainy season (Plate 6), the soil fertility status is maintained or improved by *F. albida* trees (Depommier *et al.* 1992; Sall 1992). Moreover, the deep tap root and the leaflessness during the rainy season greatly reduce the competition between *F. albida* and the crop plants for nutrients, water, and light (Schulze *et al.* 1991). Also, there is a decrease in soil temperature near *F. albida* trees which

Table 3.2–4. Dry matter production (in t ha^{-1} a^{-1}) of twigs, leaves, and wood in seven 4-year-old (1992) and 5-year-old (1993) windbreak species

	1992				1993			
Windbreak species	Twigs	Leaves	Wood	Total	Twigs	Leaves	Wood	Total
Faidherbia albida	1.3	0.3	–	1.6	1.8	0.5	1.0	3.3
Acacia holosericea	8.2	6.2	–	14.4	5.7	5.7	0.3	11.7
Azadirachta indica	3.8	1.4	2.4	7.6	4.5	2.3	5.1	11.9
Acacia nilotica	2.8	0.3	–	3.1	5.6	0.5	0.7	6.8
Andropogon gayanus	0.1	–	–	0.1	0.1	–	–	0.1
Bauhinia rufescens	4.5	0.6	1.0	6.1	6.7	0.8	1.9	9.4
Acacia senegal	8.4	0.4	–	8.8	14.1	1.8	6.4	22.3
LSD (0.05)	3.2	1.5	0.6	4.3	3.9	2.1	1.9	6.4

(*Source*: Lamers *et al.* 1994, summarized)

is closer to the optimum for millet germination and growth, thus allowing the plants to exploit the greater fertility of these plots (Vandenbeldt and Williams 1992). On the other hand, the leaflessness of this windbreak tree may provide less protection against wind erosion, as indicated by the data on plant establishment in 1992, where the number of viable millet pockets on plots with *F. albida* windbreaks was lower than on other windbreak plots. However, this decrease in millet pocket number was more than compensated by the growth stimulating effects of *F. albida.*

According to local farmers, the '*F. albida* effect' on millet production is mainly caused by the dung of free-grazing livestock which browse on the leaves and palatable fruits of *F. albida*, and search for shade during the dry season (Lamers *et al.* 1994b). Thus, under on-farm conditions, livestock may contribute greatly to soil fertility. On farmers' fields the '*F. albida* effect' may also be caused by pre-existing islands of higher soil fertility (Sanchez 1987; Geiger *et al.* 1992).

Fodder and Wood Production of the Windbreaks

Besides the protection of soils and plants from wind erosion damage, windbreaks may also provide a variety of products for agronomic and household needs (Lamers *et al.* 1994a; Plates 16, 17, 18, 19, and 20). In order to determine the agronomic and economic value of the different windbreak species, the dry matter production of twigs, leaves, and wood of the windbreak species was determined by weighing the material which was cut during the annual windbreak trimming.

The total dry matter production of the seven windbreak species varied significantly among species and from year to year (Table 3.2–4). In 1992, almost no wood (firewood) was harvested because the 4-year-old trees had not obtained the necessary height and width. But in 1993, considerable amounts of firewood were harvested on the plots with *A. indica* and *A. senegal* and smaller amounts on the plots with *B. rufescens* and *F. albida*. If the trees had been trimmed later and with longer intervals between trimmings, firewood production would probably have been higher.

The production of leaves for livestock feeding was highest on plots with *A. holosericea* followed closely by *A. indica* (Table 3.2–4). All other windbreak species produced only small amounts of leaves. However, due to the inverse life cycle and early shedding of leaves of *F. albida*, the leaf dry matter production of this tree was probably under-estimated. In terms of dry matter production, *A. holosericea* and *A. senegal* were the most productive windbreak trees and *F. albida* and *A. gayanus* the least productive. The total biomass production of *A. senegal*, *A. indica*, *A. holosericea*, and *B. rufescens* even exceeded the dry matter production of the millet crop (Table 3.2–3). However, the grass *A. gayanus* grew only slightly higher than two meters, therefore little material was cut at pruning. Thus, the results are not comparable with the data for the tree species. For *A. gayanus*, the protecting effect against wind erosion damage appears to be more important than the feed production.

The main components harvested from the windbreak trees in 1992 and 1993 were twigs and small branches that cannot be used for firewood and are generally considered to be of minor importance. However, Chase and Boudouresque (1987) reported that a mulch of twigs and branches increased the re-vegetation on bare Sahelian forest soil. Similar to the effects of crop residues (Chapters 3.3 and 3.4), the twigs and branches may also contribute significantly to the prevention of soil losses during sandstorms and 'Harmattan' winds. Local farmers are aware that twigs and branches capture considerable amounts of soil and dust and they use them to improve degraded spots on their fields (Chapter 4.3). Twigs and branches from windbreak trees may thus supplement the inadequate amounts of millet crop residues currently applied as mulching material.

Besides assessing the effects on millet grain and straw production, it is of crucial importance for the evaluation of the usefulness of windbreaks, to consider the direct fodder and wood production of the trees themselves. An economic analysis considering the multiple uses of windbreaks, as presented in Chapter 4.1, can help in selecting the appropriate windbreak-cropping system.

Conclusions

General Usefulness of Windbreaks

The data on the effect of windbreaks on wind erosion (Figs. 3.2–1 and 3.2–2) show that windbreaks may generally serve as a means to reduce soil flux and soil erosion damage in the Sahel. On the other hand, the effects of windbreaks on crop production are contradictory and depend particularly on site characteristics and windbreak management. The implementation of windbreaks in the southern Sahel should therefore be based on two factors: (1) the overall agronomic and economic benefit as described in Chapter 4.1 and (2) the ecological improvement of the landscape. With respect to the first factor, the investments for the installation of windbreaks, the loss of arable land due to windbreak plantings, and the potential decrease in crop productivity in the vicinity of the windbreaks must be correlated with the potential increase in crop yields due to protection from wind erosion and the supply of products from the windbreak trees, such as firewood, timber, fruits, or browse. With respect to the second factor, the degree to which windbreak species may prevent further soil degradation or may increase the stability of mobile sand dunes should be evaluated. Windbreaks may also help reclaim uncultivated land and increase the area of arable land. Some of these objectives may be achieved with alternative measures such as crop residue application (Chapters 3.3 and 3.4), controlled grazing, the construction of palisades with low-cost materials, or by an improved management of the existing vegetation. Nevertheless, planting trees in the Sahel could be an effective means of integrating both economic and ecological factors for wind erosion control.

Windbreak Design, Protected Area (H), and Suitable Windbreak Species

In the present study, windbreak effects on soil flux and millet productivity occurred up to a distance of ten times its height (10*H*) on plots with *F. albida* (Tables 3.2–1 and 3.2–3). For the establishment of new windbreaks, the windbreak architecture, the windbreak species, and the spacing between the windbreak rows must be considered carefully in order to achieve the maximum effect. Tall windbreak species allow a wider spacing of windbreaks but big trees may increase competition with crop plants for nutrients, water, and light, thus decreasing crop productivity, particularly close to the windbreak. The slow development of big windbreak trees also delays the harvesting of leaves and wood for fodder and firewood.

From the results of the present investigation, *F. albida* can be recommended as a suitable windbreak species for windbreak-millet cropping systems. However, this recommendation is not due to a significant effect of *F. albida* on wind erosion, but to the remarkable improvement of crop growth in its vicinity, which is likely caused by increased soil fertility. However, the slow growth of *F. albida* seedlings (Chapter 3.1) and its weak competition with weeds (Chapter 4.3) need to be considered. *B. rufescens* also appears to be a suitable tree for windbreak-millet cropping systems, because wind erosion decreased substantially (Figs. 3.2–1 and 3.2–2) and millet productivity increased slightly on these windbreak plots compared with the control. With *B. rufescens*, fodder and wood production of the windbreaks was above the average of the other windbreak species (Table 3.2–4). Despite the significant reduction in wind erosion in the vicinity of the windbreaks, *A. gayanus* cannot be recommended as a suitable windbreak species because its overall effect on wind erosion and millet dry matter production was low (Tables 3.2–1 and 3.2–3) and fodder production was minimal (Table 3.2–4).

Despite the long tradition of windbreak planting, windbreak-crop interactions are still poorly understood. Future research should focus particularly on the long-term effects of windbreaks on wind erosion (soil loss), soil fertility, crop productivity, and competition effects. Furthermore, the knowledge of local farmers and their wives on the use and management of windbreak-cropping systems should be acknowledged and included in the planning and design of new agroforestry systems (Lai 1991; Lamers *et al.* 1994a; Chapter 4.3). More information is also needed on the potential biomass production of different windbreak species as well as their requirements for light and nutrients and their optimal temperature (Maydell 1983; Sheik 1988). Technological inputs to improve tree establishment and growth, for example fertilizers and inoculation with mycorrhizal fungi, need to be evaluated. By incorporating the results of this future research into an integrated land-use concept, a better ecological and economic environment in the southern Sahel can be achieved.

Abstract

The effect of planted windbreaks on the establishment and growth of pearl millet (*Pennisetum glaucum* (L.) R. Br.) was studied in a field experiment at the ICRISAT Sahelian Center, Sadoré (Niger), during 1992 and 1993. Additionally, the biomass production for seven windbreak species was determined to evaluate the overall usefulness of windbreak-millet cropping systems as an integrated land-use concept. The windbreak trial was established in 1988. In 1992 and 1993, twelve sandstorms were counted during each of the growing seasons. Compared to the control without windbreaks, the total seasonal soil flux in the plots with *Andropogon gayanus* decreased to 90% in 1992 and to 75% in 1993, and in the plots with *Bauhinia rufescens* to 39% in 1992 and to 23% in 1993. The effect of the windbreaks on reducing soil flux decreased with increasing distance from the windbreaks. The extent of the protected area in the *A. gayanus* plots was up to five times the height (*H*) of the windbreaks (5*H*), whereas it was up to 7*H* in the *B. rufescens* plots. Despite the reduced soil flux in the windbreak plots, the number of millet pockets covered with wind-blown soil was not affected. Higher concentrations of total nitrogen, organic carbon, and exchangeable potassium, calcium, and magnesium were found in the *B. rufescens* plots at 1 m distance from the windbreaks compared to the control plots. Overall dry matter production of millet was not affected by the windbreaks. However, in *Faidherbia albida* plots, straw and grain yields were significantly higher than in any other plot, particularly at a distance less than 3 m from the windbreak.

Fodder and wood production differed significantly among the windbreak species. In 1992, leaf and twig production was highest on plots with *Acacia holosericea* and *Acacia senegal* windbreaks. The results suggest that the major importance of the windbreak-millet cropping systems in the Sahel is the increase in overall productivity and the supply of a greater variety of products for household and agricultural needs.

Résumé

Les effets de la plantation de brise-vent sur l'érosion éolienne, la croissance du petit mil, et la production de biomasse des brise-vent

Les effets de brise-vent plantés, sur l'implantation et la croissance du petit mil (*Pennisetum glaucum* (L.) R. Br.) ont été étudiés sur le terrain lors d'une expérience menée au Centre Sahélien de Sadoré (Niger), en 1992 et 1993. De plus, la production de biomasse générée en fonction des espèces de brise-vent a été déterminée pour évaluer l'utilité globale des systèmes de culture petit mil-brise-vent en tant que conception de l'usage intégré de la terre. Les essais de brise-vent ont débuté en 1988. En 1992 et en 1993 douze tempêtes de sable ont été dénombrées durant chaque période de croissance. Comparativement aux parcelles-témoins, le flux du sol total saisonnier a diminué jusqu'à atteindre 90% en 1992 et 75% en 1993 sur les parcelles à *Andropogon gayanus* et jusqu'à 39% en 1992 et de 23% en 1993 sur les parcelles à *Bauhinia rufescens.* La capacité des brise-vent à réduire le flux du sol décroît en fonction de leur éloignement. L'étendue des zones protégées dans les parcelles à *A. gayanus* était cinq fois supérieure à la hauteur (*H*) des brise-vent (5*H*) alors qu'elle était sept fois supérieure à leur hauteur (7*H*) dans les parcelles à *B. rufescens.* Malgré la réduction du flux du sol sur les parcelles à brise-vent, le nombre de poquets de petit mil couverts de sol érodé n'a pas été modifié. De plus grands taux d'azote total, de carbone organique, de potasse échangeable, de calcium et de magnésium ont été relevés sur les parcelles à *B. rufescens* à une distance de 1 m des brise-vent en comparaison avec les parcelles-témoins. La production globale de matière sèche de petit mil n'a pas été modifiée par les brise-vent. Cependant, sur les parcelles de *Faidherbia albida*, la production de paille et de grain s'est avérée nettement supérieure à celle des autres parcelles, en particulier à une distance inférieure à 3 m des brise-vent.

Le fourrage et le bois produits par les brise-vent mêmes différaient largement d'une espèce à l'autre. En 1992, les brise-vent d'*Acacia holosericea* et d'*Acacia senegal* ont produit la plus grande quantité de feuilles et de petits branchages. Les résultats suggèrent que l'importance majeure des systèmes de cultures brise-vent-petit mil au Sahel est liée à l'augmentation de la production globale, et à l'apport d'une plus grande variété de produits pour les besoins domestiques ou agricoles.

References

Bationo, A., S.H. Chien, J. Henao, B.C. Christianson, and A.U. Mokwunye. 1990. Agronomic evaluation of two unacidulated and partially acidulated phosphate rocks indigenous to Niger. *Soil Science Society of America Journal* 54: 1772–1777.

Belsky, A.J., R.G. Amundson, J.M. Duxburry, S.J. Riha, A.R. Ali, and S.M. Mwonga. 1989. The effect of trees on their physical, chemical and biological environments in a semi-arid savanna in Kenya. *Journal of Applied Ecology* 26: 1005–1024.

Ben Salem, B. 1991. Prevention and control of wind erosion in arid regions. *Unasylva* 164(42): 33–39.

Bird, P.R. 1984. Effects of trees on agricultural productivity. pp. 13–299. *In* Proceedings National Conference Focus on Farm Trees 2, University of England, Armidale, NSW, Canberra Publishing and Printing Co. Fyswick ACT.

Bird, P.R., D. Bicknell, P.A. Bulman, S.J.A. Burke, J.F. Leys, J.N. Parker, F.J. Van der Sommen, and P. Voller. 1992. The role of shelter in Australia for protecting soils, plants and livestock. *Agroforestry Systems* 20: 59–86.

Bley, J. 1990. Experimentelle und modellanalytische Studien zum Wasser- und Nährstoffhaushalt von Perlhirse (*Pennisetum americanum* L.) in Südwest-Niger. Diss. Universität Hohenheim, Stuttgart, Germany.

Brenner, A.J. 1991. Tree-crop interactions within a Sahelian windbreak system. Ph.D. Diss. University of Edinburgh, U.K.

Chase, R. and E. Boudouresque. 1987. Methods to stimulate plant regrowth on bare Sahelian forest soils in the region of Niamey, Niger. *Agriculture, Ecosystems and Environment* 18: 211–221.

Den Heijer, R.P. 1990. Trees tame the wind: The use of wind-breaks in the Sahel. *AT Source* 18(2): 9–13.

Depommier, D., E. Janodet, and R. Oliver. 1992. pp. 111–115. *In* R.J. Vandenbeldt (ed.) *Faidherbia albida* in the West African semi-arid Tropics. Proceedings Workshop ICRISAT and ICRAF, Niamey, Niger, 22–26 April 1991, ICRISAT, Patancheru, India and ICRAF, Nairobi, Kenya.

Fryrear, D.W. 1986. A field dust sampler. *Journal of Soil and Water Conservation* 41(2): 117–120.

Geiger, S.C., R.J. Vandenbeldt, and A. Manu. 1992. Pre-existing soil fertility and the variable growth of *Faidherbia albida*. pp. 121–125. *In* R.J. Vandenbeldt (ed.) *Faidherbia albida* in the West African semi-arid Tropics. Proceedings Workshop ICRISAT and ICRAF, Niamey, Niger, 22–26 April 1991, ICRISAT, Patancheru, India and ICRAF, Nairobi, Kenya.

Gorse, J.E. and D.R. Steeds. 1985. Desertification in the Sahelian and Sudanian zones of West Africa. World Bank Technical Paper No. 61, The World Bank, Washington, D.C.

Guinaudeau, C. 1987. Lutte contre l'ensablement des terres de culture au Niger, départements de Niamey, Zinder, Diffa. Consultation brise-vent, Rapport de mission (provisoire). Programme de Coopération Technique, FAO, Rome, Italy.

Hagen, L.J. and E.L. Skidmore. 1971. Windbreak drag as influenced by porosity. *Transactions of the ASAE* 14(3): 464–465.

Hue, N.V., G.R. Craddock, and F. Adams. 1986. Effects of organic acids on aluminum toxicity in subsoils. *Soil Science Society of America Journal* 50: 28–34.

Kaisin, E.F.P. 1994. An assessment of the potential for windbreaks for rainfed agricultural development in the African Sahel. pp. 214–217. *In* Proceedings 4th International Symposium on Windbreaks and Agroforestry, Viborg, Denmark, 26–30 July 1993, Hedeselskabet, Denmark.

Kerkhof, P. 1990. Agroforestry in Africa: A survey of project experience. pp. 113–123. G. Foley and G. Barnard (eds.) Panos Technical Series. Panos Publications, London, U.K.

Kessler, J.J. and H. Breman. 1991. The potential of agroforestry to increase primary production in the Sahelian and Sudanian zones of West Africa. *Agroforestry Systems* 13: 41–62.

King, K.F.S. 1989. The history of agroforestry. pp. 3–11. *In* P.K.R. Nair (ed.) Agroforestry systems in the Tropics. Kluwer Academic Publishers, Dordrecht, The Netherlands.

Lai, C.K. 1991. Design of agroforestry systems: Some examples and lessons from Bangladesh. *Forest Ecology and Management* 45: 193–198.

Lamers, J.P.A., K. Michels, B.E. Allison, R.J. Vandenbeldt, and M. von Oppen. 1994a. Trees for agroforestry systems in Niger: Researchers' gains through local knowledge of men and women. *Quarterly Journal of International Agriculture* 33(2): 179–191.

Lamers, J.P.A., K. Michels, and R.J. Vandenbeldt. 1994b. Trees and windbreaks in the Sahel: Establishment, growth, nutritive and calorific values. *Agroforestry Systems* 26(3): 171–184.

Le Houérou, H.N. 1989. The grazing land ecosystems of the African Sahel. Springer-Verlag, Berlin, Germany.

Long, S.P. and N. Persaud. 1988. Influence of neem (*Azadirachta indica*) windbreaks on millet yield, microclimate, and water use in Niger, West Africa. *In* Proceedings International Conference on Dryland Farming, Bushland/Amarillo, TX, 19–22 Aug. 1988.

Von Maydell, H.-J. 1983. Arbres et arbustres du Sahel: Leurs caractéristiques et leurs utilisations. TZ-Verlagsgesellschaft, Rossdorf, Germany.

Michels, K. 1994. Wind erosion in the southern Sahelian zone. Diss. Universität Hohenheim, Verlag Ulrich E. Grauer, Stuttgart, Germany.

Okorio, J. 1992. The effect of *Faidherbia albida* on soil properties in a semi-arid environment in Morogoro, Tanzania. pp. 117–120. *In* R.J. Vandenbeldt (ed.) *Faidherbia albida* in the West African semi-arid Tropics. Proceedings Workshop ICRISAT and ICRAF, Niamey, Niger, 22–26 April 1991, ICRISAT, Patancheru, India and ICRAF, Nairobi, Kenya.

Payne, W.A., R.J. Lascano, L.R. Hossner, C.W. Wendt, and A.B. Onken. 1991. Pearl millet growth as affected by phosphorus and water. *Agronomy Journal* 83: 942–948.

Puri, S. and K.S. Bangarwa. 1992. Effects of trees on the yield of irrigated wheat crop in semi-arid regions. *Agroforestry Systems* 20: 229–241.

Raussen, T. 1990. Planung von Windschutz-Pflanzungen in ariden und semi-ariden Gebieten. Der Tropenlandwirt, Beiheft 43. Deutsches Institut für tropische und subtropische Landwirtschaft, Witzenhausen, Universität Kassel, Germany.

Renard, C. and R.J. Vandenbeldt. 1990. Bordures d'*Andropogon gayanus* Kunth comme moyen de lutte contre l'érosion éolienne au Sahel. *Agronomie Tropicale* 45(3): 227–231.

Rosenberg, N.J. 1974. Windbreaks and shelter effect. pp. 238–264. *In* Microclimate: The biological environment. John Wiley and Sons, Chichester, U.K.

Sall, P.N. 1992. Site effects of *Acacia albida* Del. pp. 103–105. *In* R.J. Vandenbeldt (ed.) *Faidherbia albida* in West African semi-arid Tropics. Proceedings Workshop, ICRISAT and ICRAF, Niamey, Niger, 22–26 April 1991, ICRISAT, Patancheru, India and ICRAF, Nairobi, Kenya.

Sanchez, P.A. 1987. Soil productivity and sustainability in agroforestry systems. *In* H.A. Steppler and P.K.R. Nair (eds.) Agroforestry: A decade of development. International Council for Research in Agroforestry (ICRAF), Nairobi, Kenya.

Schulze, E.D., G. Gebauer, H. Ziegler, and O.L. Lange. 1991. Estimates of nitrogen fixation by trees on an aridity gradient in Namibia. *Oecologia* 88: 451–455.

Sheik, M.I. 1988. Planting and establishment of windbreaks in arid areas. *Agriculture, Ecosystems and Environment* 22/23: 405–423.

Skidmore, E.L. and L.J. Hagen. 1977. Reducing wind erosion with barriers. *Transactions of the ASAE* 20(5): 911–915.

Smith, D.M., P.G. Jarvis, and J.C.W. Odongo. 1996. Water use by Sahelian windbreak trees. pp. 115–128. *In* B. Buerkert, B.E. Allison, and M. von Oppen (eds.) Wind erosion in West Africa: The problem and its control. Proceedings International Symposium, Hohenheim, Germany, 5–7 Dec. 1994, Margraf Verlag, Weikersheim, Germany.

Sowers, F.W. and M. Issoufou. 1992. Pre-colonial agroforestry and its implications for the present: The case of the Sultanate of Damagaram, Niger. pp. 171–175. *In* R.J. Vandenbeldt (ed.) *Faidherbia albida* in the West African semi-arid Tropics. Proceedings Workshop ICRISAT and ICRAF, Niamey, Niger, 22–26 April 1991, ICRISAT, Patancheru, India and ICRAF, Nairobi, Kenya.

Sumberg, J. and M. Burke. 1991. People, trees and projects: A review of CARE's activities in West Africa. *Agroforestry Systems* 15: 65–78.

Tibke, G. 1988. Basic principles of wind erosion control. *Agriculture, Ecosystems and Environment* 22/23: 103–122.

Vandenbeldt, R.J. 1989. Agroforestry in the semi-arid Tropics: Strategies and research needs. pp. 191–200. *In* Soil, crop, and water management systems for rainfed agriculture in the Sudano-Sahelian zone. Proceedings Workshop, ICRISAT, Niamey, Niger, 7–11 Jan. 1987, ICRISAT, Patancheru, India.

Vandenbeldt, R.J. (ed.) 1992. *Faidherbia albida* in the West African semi-arid Tropics. Proceedings Workshop ICRISAT and ICRAF, Niamey, Niger, 22–26 April 1991, ICRISAT, Patancheru, India and ICRAF, Nairobi, Kenya.

Vandenbeldt, R.J. and J.H. Williams. 1992. The effect of soil surface temperature on the growth of millet in relation to the effect of *Faidherbia albida* trees. *Agricultural and Forest Meteorology* 60: 93–100.

Van Eimern, J., R. Karschon, L.A. Razumova, and G.W. Robertson. 1964. Windbreaks and shelterbelts. World Meteorological Organization, Technical Note No. 59. 188 pp.

Verinumbe, I. 1991. Agroforestry development in northeastern Nigeria. *Forest Ecology and Management* 45: 309–317.

Vierich, H.I.D. and W.A. Stoop. 1990. Changes in West African savannah agriculture in response to growing population and continuing rainfall. *Agriculture, Ecosystems and Environment* 31: 115–132.

Wilson, S.J. and R.U. Cooke. 1980. Wind erosion. pp. 217–251. *In* M.J. Kirkby and R.P.C. Morgan (eds.) Soil erosion. John Wiley and Sons, London, U.K.

Yadav, J.P., K.K. Sharma, and P. Khanna. 1993. Effects of *Acacia nilotica* on mustard crop. *Agroforestry Systems* 21: 91–98.

3.3
Anti-erosive, Soil Physical, and Nutritional Effects of Crop Residues

Andreas Buerkert, Karlheinz Michels, John P.A. Lamers,
Horst Marschner, and André Bationo

Introduction

The effects of millet crop residues on millet dry matter and grain yield have been the subject of many recent investigations (Bationo and Mokwunye 1991; Bationo *et al.* 1993; Hafner *et al.* 1993a, 1993b; Rebafka *et al.* 1994; Chapter 3.4). These effects have usually been attributed to changes in soil chemical parameters such as increases in pH and phosphorus (P) availability (Kretzschmar *et al.* 1991; Hafner *et al.* 1993a; Bationo *et al.* 1995), to increases in root growth (Kretzschmar *et al.* 1991), to improved biological N_2 fixation by associative microorganisms (Hafner *et al.* 1993b), and to the entrapment of dust (Geiger *et al.* 1992). However, a mulch of crop residues may also protect young millet from the effects of wind erosion. Compared with the bare control, a crop residue application at the rate of 2000 kg ha^{-1} reduced the amount of wind-blown soil (soil flux) by 46% in 1991 and 48% in 1992 (Chapter 3.4). Coverage of millet seedlings by wind-blown soil severely hampered millet establishment, reduced subsequent growth, and decreased grain yield by almost 50% compared with uncovered plants (Chapter 2.2). The objective of this study was to determine to what degree the effects of mulched crop residues on millet establishment and yields may be attributed to: (i) changes in soil chemical properties caused by decomposing residues, (ii) changes in temperature and mechanical resistance of the surface soil, and (iii) physical protection of young millet plants from the damaging effects of sandstorms.

B. Buerkert, B. E. Allison and M. von Oppen (eds.), Wind Erosion in Niger, 123–138.

Experimental Setup

Rainfed millet was planted at the station of the ICRISAT Sahelian Center at Sadoré (Niger) from 1992 to 1994 on a reddish acid sandy soil classified as an Arenosol (West *et al.* 1984; Chapter 1.2). Given the long-term average of 560 mm, annual rainfalls of 586 mm in 1992 and 542 mm in 1993 were normal, but rainfall was unusually high in 1994 (780 mm). Previously, the site had been cultivated with millet and then left fallow for two years.

In early May of 1992, nine plots each 19 by 25 m were established side by side. Given that most storms come from the east, the 25 m length was oriented east-west (Fig. 3.3–1). There were three treatments: a bare control, millet crop residues applied as surface mulch at a rate of 2000 kg ha^{-1}, and surface applied polyethylene (plastic) tubes of the length, diameter, and number similar to the crop residues. The orange tubes were cut into pieces 2 m long and partially filled with soil to make them of similar weight to the millet residues and to prevent them from being blown away. Plastic tubes were applied only once, at the onset of the 1992 rainy season. The crop residue mulch, however, was reapplied every year at the onset of the dry season after removing the leftover mulch, not decomposed by termites or microbial activity, from the previous season. Because of a 1% slope from south to north, treatments were arranged in blocks and replicated three times. In all years, all plots received a broadcast application of 13 kg P ha^{-1} as single superphosphate before the rainy season. Millet *cv.* CIVT was sown at the onset of the rainy season on 26 May 1992, 6 June 1993, and 5 June 1994 at 10 000 pockets ha^{-1}. Each pocket consisted of a 0.05 m deep hole made with a hand hoe into which 50 to 100 seeds were placed and covered with soil. Millet was thinned to three plants per pocket 25 days after sowing (DAS). Calcium ammonium nitrate at 30 kg N ha^{-1} was split into two equal applications and incorporated 0.1 m from the pockets at thinning and at booting (45 DAS). Two applications of Decis® (deltamethrine) at the rate of 1 kg ha^{-1} were used during flowering to protect the crop against damage from insect attacks, mainly the cotton stainer (*Dysdercus voelkeri* Schmidt). Weeding was done manually at thinning and booting. In this paper, only data on millet growth from maturity (100 DAS) in 1992, and thinning (25 DAS), booting (45 DAS), and maturity in 1993 are reported. Plant samples were taken as follows: at thinning 100 seedlings from each plot were selected at random; rows of millet, each comprising 23 pockets, were harvested at booting (one row) and at final harvest (15 rows). To determine dry matter, all plant material was dried to constant weight at 65°C. At maturity, millet heads were threshed by hand and grain yield determined.

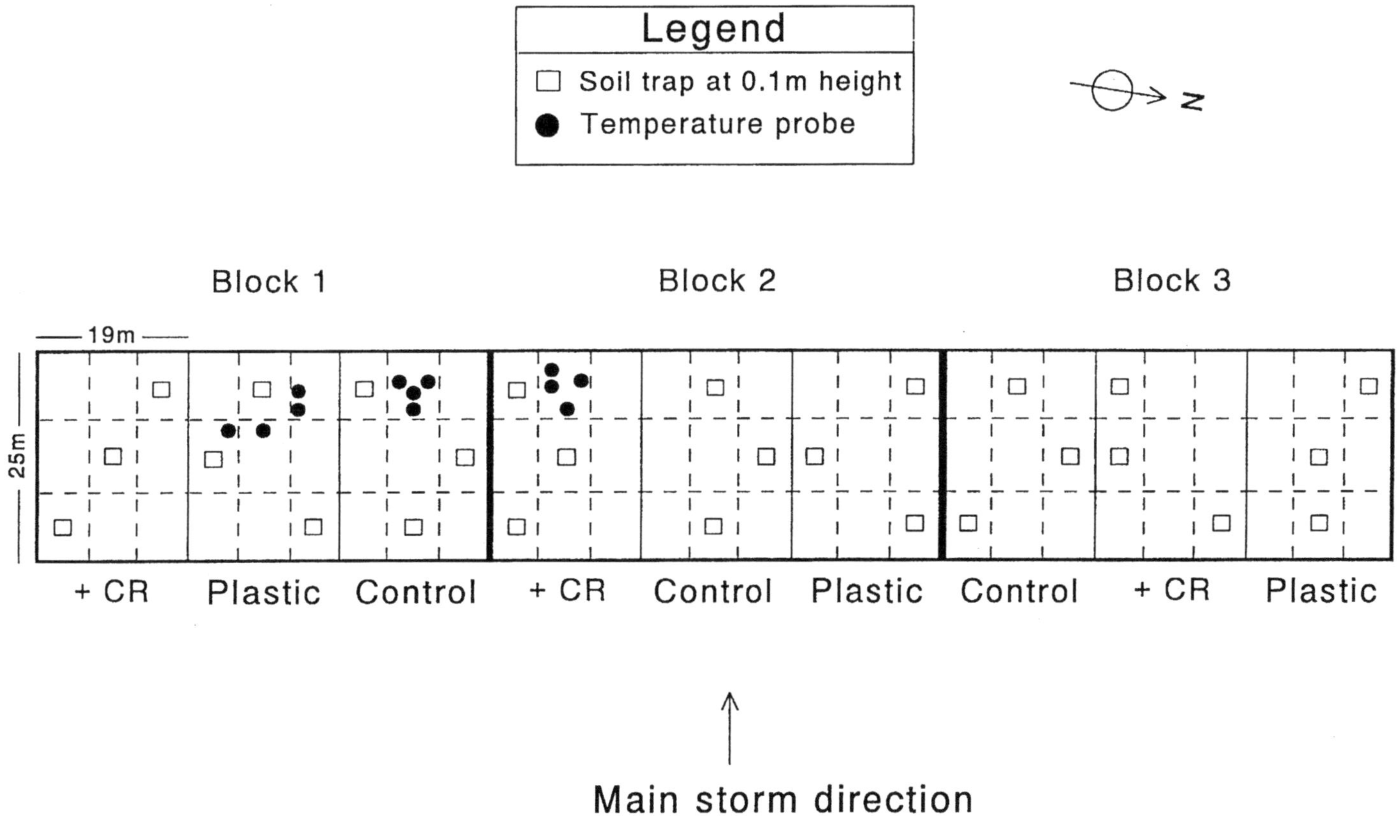

Fig. 3.3–1. Layout of the field trial at the ICRISAT Sahelian Center, Sadoré, Niger. (□) indicates the position of BSNE soil traps and (•) the position of soil temperature probes and thermocouples connected to a data logger.

Measurements of Wind Erosion and Soil Properties

Soil flux was measured with BSNE traps (Fryrear 1986; Plate 12) by placing three traps in each plot 0.1 m above ground level (Fig. 3.3–1). The traps with a vertical sampler opening of 0.001 m^2 were placed 4, 12, and 20 m from the eastern border. During the rainy season traps were emptied after each storm, and during the dry season every two weeks. To compare soil flux with changes in surface elevations and to distinguish between soil erosion and deposition, twelve U-shaped bars, 0.35 m in height and 0.2 m in width, made of stainless steel were placed in each plot at the onset of the rainy season in May 1993. This method was similar to that described by Wilson and Cooke (1980). The U-shaped bars were placed upright, cemented into holes and used as a fixed reference height which was measured before the first storm. Subsequently, the distance from the top of the bar to the soil surface was measured to the nearest millimeter. Thus, at each measurement location, a reduction in height meant 'deposition' and an increase 'erosion'. Subsamples in each plot were averaged for the three replicates. Measurements were taken twice a month during the 'Harmattan' season and after each major storm during the rainy season.

To examine treatment effects on the mechanical resistance of the surface soil, penetrometer measurements were taken before the onset of the rains in April 1993. At 40 randomly chosen locations in each plot, resistances were independently measured at 0 to 0.02 m and 0 to 0.05 m soil depth. The hand-held penetrometer (Eijkelkamp, The Netherlands) was equipped with tips 35 mm in diameter for the upper and 15 mm for the lower depth.

To measure soil surface temperatures, three temperature probes (Type 108, Campbell Scientific, England) were randomly placed at 0.01 m depth in each of the three treatments, leading to three subsamples per treatment (Fig. 3.3–1). Additionally, copper-constantan thermocouple wire was used to measure soil temperature at the 0.05 m soil depth in each of the three plots. The differential measurements of the nine temperature probes and the three thermocouple wires were recorded every 20 s, averaged every 30 min, and stored with a CR10 datalogger (Campbell Scientific, England).

In April 1994, ten soil samples were taken again in each plot to determine the chemical soil properties. The sampling depths were 0 to 0.1 m and 0.1 to 0.2 m. Soil samples were analyzed according to standard procedures for pH (1:2.5 0.01 *M* KCL), organic carbon (C) (Walkley and Black 1934), total N (Brenner and Mulvaney 1982), exchangeable aluminium (Al) and total acidity (Mc Lean 1982), and P_{Bray} (Olsen and Sommers 1982). Exchangeable bases were extracted with 1 *M* NH_4-acetate. Based on a series of test runs with different durations of incubation, the standard procedure to determine water soluble P was modified by incubating the soil for only 5 min before P readings were taken (Buerkert 1995). Thus, only those P fractions which were weakly bound and likely to be of special importance for growing plant roots, were detected.

Statistical Analyses

To test for a 'distance from the east' effect, data of the soil fluxes from the three east-west positions in each plot were considered as repeated measurements over space and analyzed with the General Linear Model procedure provided within the SAS software package (Littell *et al.* 1991). In contrast, sampling times within each season were analyzed as repeated measurements over time. Dry matter and yield data were analyzed with GENSTAT 5 release 3 (Lawes Agricultural Trust 1993) using a randomized complete block model. The measurements of soil mechanical resistance were analyzed with a split-plot model for the two depths.

Soil Flux, Erosion, Surface Resistance, and Soil Temperature

The analysis did not reveal any effect of the position of the soil traps on the 'distance from the east', subsequently subsamples for each plot were averaged. Mulching reduced total soil flux during the rainy season of 1993 by 59% and 1994 by 26% (Table 3.3–1A and C). These data are similar to the results obtained in 1991 and 1992 (Chapter 3.4). Plastic mulch was slightly more effective at reducing soil flux than crop residues, most likely due to a higher resistance against decomposition by termites. There were significant storm by treatment interactions indicating that the effectiveness of mulch in reducing soil flux depended on the severity of the storm. Total soil flux during the 1994 'Harmattan' season was about half that during the rainy season of 1993, and mulching reduced it again by 62% (Table 3.3–1B). The significant effect of mulching during the 'Harmattan' season indicates that even at this time of the year a large proportion of soil particles moving across a millet field may be attributed to short rather than long-distance transport. Unfortunately, there are no analyses available which would determine the relative proportions of particle sizes of the eroded soil material compared with the surrounding native soil material.

Averaged across treatments, the total soil flux during the 1994 rainy season was 62% greater than in 1993 and, in contrast to 1993, mulching was less effective, regardless of the amount of soil transported in a storm (Table 3.3–1C). This may be due in part to the much higher rainfall in 1994 which caused splash movement of soil particles, or to differences in storm intensity.

The measurements of the soil surface level showed that during the 1993 rainy season, erosion occurred even in mulched treatments. Visual observations indicated that soil erosion was caused by both wind and water. Mulching, however, had a distinct effect (Fig. 3.3–2). Similar to the soil flux, plots with

Table 3.3–1. Soil flux as affected by plastic mulch treatments and crop residue mulch (CR) at 0.1 m height above-ground in 1993 and 1994

Soil flux (kg m^{-2} vertical sampler opening)

A	Date of storm during rainy season 1993														Total per season
	April	May			June					July					
	27	23	30	31	06	13	16	22	24	01	06	07	10	23	
TREATMENT															
Control	17	97	10	25	18	154	13	1	16	63	95	4	3	35	551
Plastic$_{2000}$	1	17	3	13	12	64	3	1	12	28	35	3	3	8	203
CR$_{2000}$	1	38	3	12	10	64	5	1	12	34	60	3	3	8	254

B	Sampling during 'Harmattan' season 1994							
	February		March		April			
	01	15	01	15	01	15	20	
TREATMENT								
Control	20	1	7	4	42	47	129	250
Plastic$_{2000}$	1	0	1	1	15	18	72	108
CR$_{2000}$	1	0	1	1	6	7	67	83

C	Date of storm during rainy season 1994										
	May			June				July			
	02	14	25	07	10	15	23	04	07	11	
TREATMENT											
Control	26	66	144	37	48	173	44	79	2	36	655
Plastic$_{2000}$	4	36	116	27	50	144	31	54	2	25	489
CR$_{2000}$	9	36	111	15	41	154	38	49	2	31	486

the plastic mulch had the least erosion. With the onset of the 'Harmattan' winds in February 1994, the surface level of mulched plots increased by almost 7 mm with crop residues and 4 mm with the plastic mulch application. This corresponds to a soil deposition of 100 and 60 t ha^{-1} in crop residue and plastic plots, respectively. Herrmann *et al.* (1994) estimated the amount of dust deposited during the 'Harmattan' season to be 1.9 t ha^{-1}, which again indicates that the deposition measured in our trial was largely due to short distance transportation of soil particles. On the bare soil, where deposition was negligible, any fallout from the air was probably retranslocated and eventually moved out of the experimental area (Plate 21). In contrast to the 1993 rainy season, the onset of the 1994 rains resulted in increased rates of erosion only in the bare plots, whereas soil deposition occurred in plots with plastic and crop residue mulch (Fig. 3.3–2).

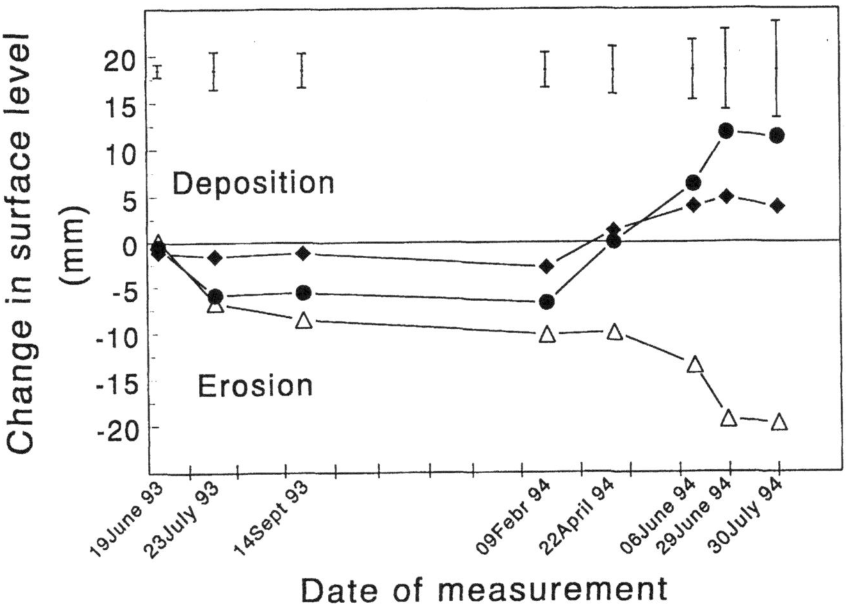

Fig. 3.3–2. Soil erosion and deposition from June 1993 to July 1994. (△) indicates the bare control, (●) a mulch of 2000 kg ha^{-1} of crop residues (CR_{2000}), and (u) a mulch of plastic tubes ($Plastic_{2000}$). Data points represent means of three replicates each of which is based on twelve subsamples. Error bars indicate twice the standard error of the difference.

Comparing years, the results of the erosion and deposition measurements are in contrast to the soil flux data. Mulch effects were much smaller in 1994 than in 1993. The reasons for this are not clear but it may be caused, in part, by differences in surface runoff between years. This is an indication that soil flux measurements should be used with caution when drawing conclusions about soil erosion. Deposition in the plastic mulch treatment was smaller than in the crop residue treatment (Fig. 3.3–2). This most likely occurred because by 1994, the plastic mulch had become less effective due to continuous exposure to UV radiation and to attacks by termites. In the crop residue plots fresh material had been placed every year and led to increased surface roughness. In 1994, two months of heavy sand- and rainstorms, from late May to late July, decreased the soil surface level in bare plots by 9 mm, which is equivalent to a soil erosion loss of about 140 t ha^{-1}. A similar amount, 180 t ha^{-1}, was deposited in plots that were mulched with 2000 kg ha^{-1} of crop residues. Soil crusts (Hoogmoed and Stroosnijder 1984) may have formed quickly on the control plots hindering infiltration and leading to increased surface runoff. However, the fact that erosion in bare plots and deposition in mulched plots was similar in total amounts, indicates that wind erosion must have played a predominant role in the processes involved. The amount of soil eroded

Table 3.3–2. Mechanical resistance of the surface soil at 0 to 0.02 m and 0 to 0.05 m depth measured with a manually operated penetrometer in 1993 and 1994, as affected by plastic mulch treatment and crop residue mulch (CR)

	Mechanical resistance (kN m^{-2})	
	[0 to 0.02 m depth]	[0 to 0.05 m depth]
Treatment		
May 1993		
Control	283	2400
$Plastic_{2000}$	193	1490
CR_{2000}	174	1380
sed	*46*	*350*
May 1994		
Control	137	2150
$Plastic_{2000}$	73	1570
CR_{2000}	67	1310
sed	*14*	*330*

sed = standard error of the difference

and accumulated is very similar to that estimated by Geiger *et al.* (1992) at the ICRISAT Sahelian Center. The authors reported an absolute difference in the surface level between bare and mulched plots of 0.15 m after five years of mulching. This is one-third larger than the difference of 0.02 m a^{-1} determined in this study.

A surface cover with either type of mulch decreased the mechanical resistance of the crusted soil surface at 0 to 0.02 m and 0 to 0.05 m depth (Table 3.3–2). The effect that covering a surface with organic material had on soil resistance was primarily on increased activities of soil flora and fauna, but in particular of termites. Termites also attacked the plastic mulch, which is reflected in the lower surface soil resistance in plots with polyethylene tubes compared with the control in both years.

Mulching also had significant effects on temperatures of the surface soil (Fig. 3.3–3). The application of crop residues decreased the maximum daily temperature at 0.01 m soil depth by about 8°C. Since the percentage of soil covered with the mulch was estimated to be only about 10% (Michels *et al.* 1995), temperature effects from the mulch are more likely due to the storage of moisture in and around the mulching material in the soil rather than to shading effects. With the extremely high surface soil temperatures in the Sahel (Lawson and Sivakumar 1991; Vandenbeldt and Williams 1992), the temperature buffering effect from the mulch could be decisive for the survival

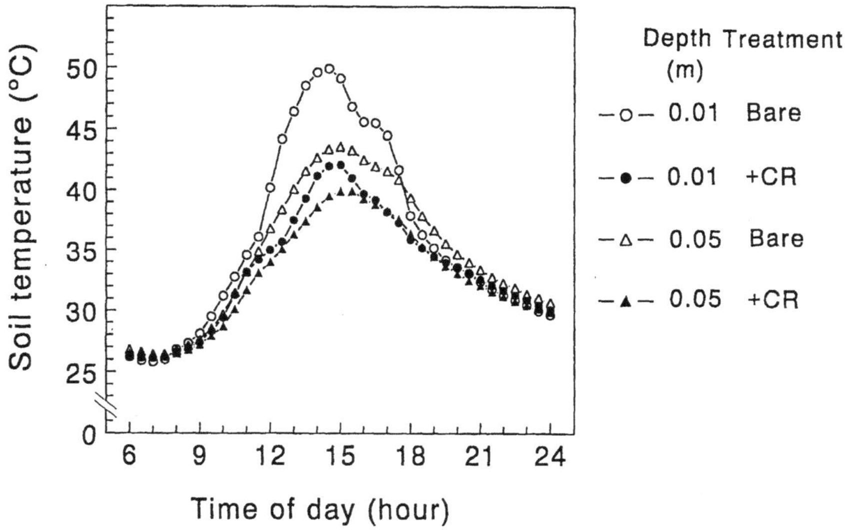

Fig. 3.3–3. Diurnal pattern of surface soil temperatures in bare and crop residue (CR) mulched plots at 0.01 m and 0.05 m depth on a typical clear day after the onset of the rainy season in June 1993. Data points at the 0.05 m depth are means of three subsamples per treatment. At the 0.01 m depth, however, there was only one thermocouple per treatment.

of millet plants, especially when the millet seedlings are partially covered by soil and have their meristems in direct contact with the hot surface soil. The effect of mulching on soil temperature decreased with depth and was also smaller with the plastic mulch than with the crop residue mulch.

Soil Chemical Properties

After two years, both types of surface mulches had caused significant changes in the chemical properties of the surface soil (Table 3.3–3). Mulching increased the pH and, consequently, decreased exchangeable Al concentration at 0 to 0.1 m and 0.1 to 0.2 m soil depth. Exchangeable cations were clearly higher in the mulched plots compared with the control. Similar results were found in long-term studies with crop residue mulch applications by Bationo *et al.* (1995). With a crop residue mulch, these increases were due to both the amount of nutrients in the soil material deposited by sandstorms and to the nutrients supplied by decomposing millet stalks.

Rebafka *et al.* (1994) showed that a long-term removal of crop residues on the acid sandy soils of the Sahel may lead to potassium (K) deficiency, given the high K uptake of millet. Therefore, the amount of K that was leached into the surface soil from the crop residue mulch was calculated. Assuming a bulk density of 1.65 g cm^{-3}, the calculations based on the data of Table 3.3–3

Table 3.3–3. Soil chemical properties at 0 to 0.1 m and 0.1 to 0.2 m soil depth after two years of mulching (April 1994)

Treatment	Organic C (%)	N_{total} (mg kg^{-1})	P_{Bray} (mg kg^{-1})	P_{H_2O} (μg kg^{-1})	pH	Al^{3+} (cmol$_c$ kg^{-1})	K^+ (cmol$_c$ kg^{-1})	Ca^{2+} (cmol$_c$ kg^{-1})	CEC (cmol$_c$ kg^{-1})
0 to 0.1 m depth									
Control	0.16	144	13	157	4.3	0.31	0.06	0.24	0.88
$Plastic_{2000}$	0.18	161	18	253	4.5	0.16	0.07	0.36	0.84
CR_{2000}	0.17	151	18	440	4.6	0.11	0.11	0.45	0.92
sed	*0.01*	*19*	*1*	*42*	*0.06*	*0.04*	*0.004*	*0.04*	*0.04*
0.1 to 0.2 m depth									
Control	0.14	131	8	157	4.3	0.42	0.06	0.18	0.95
$Plastic_{2000}$	0.14	127	12	230	4.4	0.26	0.07	0.26	0.85
CR_{2000}	0.14	129	10	207	4.4	0.27	0.08	0.24	0.85
sed	*0.01*	*14*	*1*	*42*	*0.05*	*0.05*	*0.004*	*0.02*	*0.05*

sed = standard error of the difference

show that at the 0 to 0.1 m soil depth, the difference in exchangeable K between crop residue and plastic mulch is approximately 12.9 kg K ha^{-1} a^{-1} and at 0.1 to 0.2 m soil depth 3.2 kg K ha^{-1} a^{-1}. At an annual crop residue application rate of 2000 kg ha^{-1} and a K concentration in the straw material of about 15 mg K g^{-1}, a maximum of 30 kg K ha^{-1} a^{-1} could have been provided. However, the 16.1 kg K ha^{-1} a^{-1} in the surface soil reflected a recovery rate of only 54%. A fixation of K is unlikely, given the low clay content of the soil (Chapter 3.4). The difference, therefore, may be due to the leaching of K into deeper soil layers and also to the proportion of the K in the partly decomposed millet stalks that was removed at the end of the growing season when fresh material was applied to the plots. The slight increase in K concentration of the surface soil covered with the plastic mulch may have been a result of K deposited with the wind-blown soil material. However, the total annual K deposit was only about 6.4 kg K ha^{-1}, which is less than half of the amount predicted from dust collections on fallow land (Herrmann *et al.* 1993).

The soil chemical analyses suggest that annually 14.9 kg Ca ha^{-1} and 4.0 kg Mg ha^{-1} originated from decomposing crop residues (Table 3.3–3). Given the release of 5.1 kg Ca ha^{-1} a^{-1} and 4.9 kg Mg ha^{-1} a^{-1} from millet stalks, this would translate into recovery rates of 292% for Ca and 82% for Mg. Dust deposition increased the amount of Ca in the surface soil annually by 33 kg ha^{-1} and of Mg by 4.0 kg ha^{-1}. This compares with 9.3 kg

Ca ha^{-1} a^{-1} and 5.2 kg Mg ha^{-1} a^{-1} estimated from dust collections by Herrmann *et al.* (1993). The very high increases of Ca in mulched plots can only be explained by the fate of the 20% Ca contained in the 165 kg ha^{-1} of single superphosphate applied to all plots annually. It was observed that large amounts of the light gypsum granules remaining from the fertilizer were washed from the bare control plots with the heavy rains in July and August, but there appeared to be only small losses of granules in the mulched plots. Similarly, the extra amount of Ca in the crop residue plots compared with the plastic plots cannot be explained by leaching from the millet stalks, but suggests that crop residues were slightly more effective in reducing runoff than the plastic mulch. This difference in runoff may be attributed to the larger water absorption of the organic mulch material and to its closer contact with the soil, which led to slightly larger microridges of wind-blown soil particles around millet stalks compared with the plastic tubes.

Irrespective of treatment, changes in soil organic matter such as those described in Chapter 3.4 could not be detected (Table 3.3–3). In contrast to findings by Michels *et al.* (1995), this confirms the results of Bationo *et al.* (1995) showing that even a long-term crop residue application cannot increase but only maintain the soil organic matter content on the acid sandy soils of the Sahel. Both types of mulch increased extractable P, as determined by P_{Bray} by 40%, whereas water extractable P increased by 56% with plastic mulch and by 175% with crop residue mulch. These increases may be attributed to the changes in pH and exchangeable Al but, for crop residues, possibly also to organic acids which may have leached from the decomposing millet straw and displaced P at sorption sites (Hue 1991).

Dry Matter and Grain Yield of Millet

At maturity in 1992, straw and grain yields of millet were similar for the bare control and the plastic mulch, but they were 39 and 31% higher with the crop residue application (Table 3.3–4). Both types of mulch were similarly effective in increasing stand density by protecting millet from erosion and plant coverage with soil material. Thus, the effect of crop residues on millet straw and grain yields at maturity was due to both a 5% higher final plant density and a 29% increase in total dry matter yield per pocket (Table 3.3–4). The results also indicate that the effect of a crop residue mulch on grain yield was largely due to an increased number of heads per pocket, which may reflect higher nutrient availability.

Compared with the control in 1993, mulching led to an increase in dry matter at thinning, 30% with a plastic mulch and 60% with crop residue application (Table 3.3–5). At booting, the 260% dry matter increase with a crop residue mulch compared with the control was caused by a 15% increase in plant density, and by a threefold increase in dry matter yield per pocket

Table 3.3–4. Plant stand, heads per pocket, and dry matter yield of millet at maturity in 1992 as affected by plastic mulch treatments and crop residue mulch (CR)

Treatment	Pockets per ha	Heads per pocket	Dry matter Straw	Grain	Total	Total
			—— ($kg\ ha^{-1}$) ——			(g per pocket)
Control	9330	3.6	1240	740	2360	253
$Plastic_{2000}$	9770	3.4	1320	780	2490	255
CR_{2000}	9750	3.8	1720	970	3180	326
sed	*320*	*0.3*	*330*	*210*	*630*	*66*

sed = standard error of the difference

Table 3.3–5. Plant stand and dry matter yield of millet at thinning and booting in 1993 as affected by plastic mulch treatments and crop residue mulch (CR)

Treatment	Pockets per ha	Dry matter Thinning ($kg\ ha^{-1}$)	Booting ($kg\ ha^{-1}$)	(g per pocket)
Control	8700	1.0	37	4.3
$Plastic_{2000}$	9860	1.3	67	6.8
CR_{2000}	10 000	1.6	132	12.2
sed	*520*	*0.3*	*31*	*3.1*

sed = standard error of the difference

(Table 3.3–5). This effect lasted until maturity when most of the total dry matter increase per hectare in crop residue plots came from a 71% increase in dry matter yield per pocket. In both study years, the two types of mulch equally increased final stand density, but the effects were larger in 1993 with stronger erosion by wind or water than in 1992 (Table 3.3–6). Also in 1993, in contrast to the previous year, the plastic mulch increased straw yield by 67% and grain yield by 64%. The effects of the chemically inert polyethylene material on millet growth from thinning to maturity can be, in part, explained by the mulching effects on plant stand. These yield increases seem to also reflect the chemical and physical changes in surface soil characteristics caused by the plastic mulch.

The rate of 2000 kg crop residues ha^{-1} applied as a surface mulch in our trial is rather high compared with the amounts left on farmers' fields at the onset of the rainy season. Currently, farmers in Niger seldom broadcast crop residues on their entire fields (Plate 22). Reasons for this are: the low levels

Table 3.3–6. Plant stand, heads per pocket, and dry matter yield of millet at maturity in 1993 as affected by plastic mulch treatments and crop residue mulch (CR)

Treatment	Pockets per ha	Heads per pocket	Dry matter Straw	Grain	Total	Total
			—— ($kg\ ha^{-1}$)		——	(g per pocket)
Control	7910	2.2	548	232	1024	129
$Plastic_{2000}$	9790	2.7	917	380	1478	151
CR_{2000}	9640	3.4	1223	479	2128	221
sed	*890*	*0.4*	*210*	*69*	*333*	*35*

sed = standard error of the difference

of millet stalks produced without the application of mineral fertilizers or manure, their competing multiple uses (Lamers and Feil 1993; Chapter 4.3), the increased weed growth after crop residue application, and the relatively low effectiveness of mulched crop residues (in terms of yield increases) on productive parts of their fields (Buerkert *et al.* 1995). Therefore in the context of an entire farming system, the positive effects of mulched crop residues described in this chapter are limited to small spots on a farmer's cropland where high amounts of mulching material are applied for the purpose of soil reclamation (Chapter 4.3; Plate 23). It is only with the application of mineral fertilizers that the increase in crop biomass at the farm level would be large enough to allow a broadcast mulch application on entire fields (Chapter 4.2). However under farmers' conditions, erosion on bare spots may be also limited by the proximity of trees, bushes, strips of fallow land, grass swards or even weeds, which serve as natural micro-windbreaks or soil coverage.

Conclusions

Undoubtedly, one reason for the increases in millet yields after the application of mulched crop residues is changes in soil chemical properties which lead to a better availability of nutrients, mainly P. Additionally, crop residues reduced the abrasive soil flux, coverage by wind-blown soil, and the surface soil temperature. Mulched crop residues also improve plant establishment and dry matter production of millet under the agro-climatic conditions that prevail in the Sahel. These beneficial physical effects are complemented by a decrease in the mechanical resistance of the surface soil, which may lead to better water infiltration and increases in root growth of millet after crop residue applications. The relative importance of chemical *versus* physical effects of crop residues on millet growth depends on the soil's susceptibility to wind erosion and the crop growth stage at which damage occurs.

Abstract

In the Sahelian region of West Africa, increases in pearl millet (*Pennisetum glaucum* (L.) R. Br.) yields after a crop residue application have been attributed primarily to changes in soil chemical properties. These yield increases, however, may also be due to physical effects such as protection of millet seedlings from sandstorms. Therefore, the relative importance of soil chemical *versus* mechanical effects was investigated from 1992 to 1994 on an Arenosol in southern Niger. The application of mulched millet crop residues was compared with inert polyethylene (plastic) tubes of length, weight, and diameter similar to millet residues. Soil flux, soil erosion, mechanical resistance of the surface soil, and mid-day surface soil temperatures were significantly lower in both mulching treatments compared with the control without mulch application. Depending on the year, above-ground millet dry matter at maturity in plots covered with 2000 kg ha^{-1} of crop residues was 35 to 109% and in plastic mulch plots 6 to 44% higher than in the bare control plots. Final stand density was similar with both types of mulch and ranged from 5 to 23% higher than in the bare control. Polyethylene tubes were as effective as crop residues in reducing the effects of wind erosion on pearl millet seedlings. Crop residues not only protected seedlings from erosion damage but they also increased dry matter per plant. This was, in part, due to an improved nutrient supply.

Résumé

Effets anti-érosifs et effets physiques et nutritionnels des résidus de récoltes sur le sol

Dans les régions sahéliennes d'Afrique de l'Ouest, les augmentations de production de petit mil (*Pennisetum glaucum* (L.) R. Br.) après application de résidus de récoltes ont tout d'abord été attribuées aux changements des caractéristiques chimiques du sol. Ces augmentations de production peuvent cependant être dûes aussi à des effets physiques tels que la protection des plantules de petit mil des tempêtes de sable. C'est pourquoi on a étudié l'importance relative des effets chimiques par rapport aux effets physiques sur un Arénosol du sud Niger de 1992 à 1994. On a comparé les applications de résidus de récoltes de paillis de petit mil d'une part et de tubes de polyéthylène inerte (plastique) d'autre part, de taille, poids et diamètre similaires à ces résidus. Le flux du sol, son érosion, ainsi que sa résistance mécanique et les températures de mi-journée en surface étaient nettement inférieurs sur les parcelles soumises au mulching que sur les parcelles-témoins. Selon l'année, à maturité, la matière sèche récoltée des parties aériennes du petit mil était supérieure de 35 à 109% sur les parcelles recouvertes de 2000 kg ha^{-1} de résidus de récoltes, par rapport aux parcelles-témoins nues, alors que sur les parcelles soumises au mulching de plastique, la matière sèche avait augmenté de 6 à 44%. La densité finale en pieds était similaire dans les deux cas de mulching et de 5 à 23% supérieure que sur les parcelle-témoins. Les tubes de polyéthylène ont été aussi efficaces que les résidus de récoltes à réduire les effets de l'érosion éolienne sur les semis de petit mil. Les résidus de récoltes ont cependant non seulement protégé les semis des effets néfastes de l'érosion mais ont également augmenté la quantité de matière sèche par plant. Ceci est dû en partie à une amélioration des apports nutritifs.

References

Bationo, A., C.B. Christianson, and M.C. Klaij. 1993. The effect of crop residue and fertilizer use on pearl millet yields in Niger. *Fertilizer Research* 34: 251–258.

Bationo, A., A. Buerkert, M.P. Sedogo, B.C. Christianson, and A.U. Mokwunye. 1995. A critical review of crop residue use as soil amendment in the West African semi-arid Tropics. pp. 305–322. *In* M. Powell, S. Fernández, T.O. Williams, and C. Renard (eds.) Livestock and sustainable nutrient cycling in mixed farming systems of sub-Saharan Africa. II. Technical papers. Proceedings International Conference, Addis Ababa, Ethiopia, 22–26 Nov. 1993, ILCA (International Livestock Centre for Africa), Addis Ababa, Ethiopia.

Bationo, A. and A.U. Mokwunye. 1991. The role of manure and crop residues in alleviating soil fertility constraints to crop production: With special reference to the Sahelian and Sudanian zones of West Africa. *Fertilizer Research* 29: 117–125.

Brenner, J.M. and C. S. Mulvaney. 1982. Nitrogen – total. pp. 595–624. *In* A.L. Page *et al.* (eds.) Methods of soil analysis. Part 2. Chemical and microbial properties. 2nd ed. Agronomy: A Series of Monographs 9, ASA and SSSA, Madison, WI.

Buerkert, A. 1995. Effects of crop residues, phosphorus and spatial variability on yield and nutrient uptake of pearl millet (*Pennisetum glaucum* L.) in southwest Niger. Diss. Universität Hohenheim, Verlag Ulrich E. Grauer, Stuttgart, Germany.

Buerkert, A., R.D. Stern, and H. Marschner. 1995. Post stratification clarifies treatment effects on millet growth in the Sahel. *Agronomy Journal* 87: 752–761.

Fryrear, D.W. 1986. A field dust sampler. *Journal of Soil and Water Conservation* 41(2): 117–120.

Geiger, S.C., A. Manu, and A. Bationo. 1992. Changes in a sandy Sahelian soil following crop residue and fertilizer additions. *Soil Science Society of America Journal* 56: 172–177.

Hafner, H., J. Bley, A. Bationo, P. Martin, and H. Marschner. 1993a. Long-term nitrogen balance for pearl millet (*Pennisetum glaucum* L.) in an acid sandy soil of Niger. *Zeitschrift Pflanzenernaehrung und Bodenkunde* 156: 164–176.

Hafner, H., E. George, A. Bationo, and H. Marschner. 1993b. Effect of crop residues on root growth and nutrient acquisition of pearl millet in an acid sandy soil in Niger. *Plant and Soil* 150: 117–127.

Herrmann, L., K.E. Bleich, and K. Stahr. 1993. Veränderung eines landwirtschaftlich genutzten Standortes in Niger/Westafrika durch Staubeintrag. *Mitteilungen der Deutschen Bodenkundlichen Gesellschaft* 72: 927–930.

Herrmann, L., K.E. Bleich, and K. Stahr. 1994. Influence des dépôts éoliens sur la restauration de la fertilité des sols sous végétation en friche au Niger, Afrique de l'Ouest. *Réseau Erosion Bulletin* 14: 74–81.

Hoogmoed, W.B. and L. Stroosnijder. 1984. Crust formation on sandy soils of the Sahel. I. Rainfall and infiltration. *Soil and Tillage Research* 4: 5–23.

Hue, N.V. 1991. Effects of organic acids/anions on P sorption and phytoavailability in soils with different mineralogies. *Soil Science* 152(6): 463–471.

Kretzschmar, R.M., H. Hafner, A. Bationo, and H. Marschner. 1991. Long and short-term effects of crop residues on aluminium toxicity, phosphorus availability and growth of pearl millet in an acid sandy soil. *Plant and Soil* 136: 215–223.

Lamers, J. and P. Feil. 1993. The many uses of millet residues. *ILEIA Newsletter* 9: 15.

Lawes Agricultural Trust. 1993. GENSTAT 5 Release 3 Reference Manual. Oxford University Press, Oxford, U.K.

Lawson T.L. and M.V.K. Sivakumar. 1991. Climatic constraints to crop production and fertilizer use. *Fertilizer Research* 29: 9–20.

Littell, R.C., R.J. Freud, and P.C. Spector. 1991. SAS system for linear models. 3rd ed. SAS Inst. Inc., Cary, NC.

McLean, E.O. 1982. Soil pH and lime requirement. pp. 199–224. *In* A.L. Page *et al.* (eds.) Methods of soil analysis. Part 2. Chemical and microbial properties. 2nd ed. Agronomy: A Series of Monographs 9, ASA and SSSA, Madison, WI.

Michels, K., M.V.K Sivakumar, and B.E. Allison. 1995. Wind erosion control using crop residues: I. Effects on soil flux and chemical properties. *Field Crops Research* 40(2): 101–110.

Olsen, S.R. and L.E. Sommers. 1982. Phosphorus. pp. 403–430. *In* A.L. Page *et al.* (eds.) Methods of soil analysis. Part 2. Chemical and microbial properties. 2nd ed. Agronomy: A Series of Monographs 9, ASA and SSSA, Madison, WI.
Rebafka, F.P., A. Hebel, A. Bationo, K. Stahr, and H. Marschner. 1994. Short and long-term effects of crop residues and of phosphorus fertilization on pearl millet yield on an acid sandy soil in Niger, West Africa. *Field Crops Research* 36: 113–124.
Vandenbeldt, R.J. and J.H. Williams. 1992. The effect of soil surface temperature on the growth of millet in relation to the effect of *Faidherbia albida* trees. *Agricultural and Forest Meteorology* 60: 93–100.
Walkely, A. and I.A. Black. 1934. An examination of the Degtjareff methods for determining soil organic matter and a proposed modification of the chromic acid titration method. *Soil Science* 37: 29–38.
West, L.T., L.P. Wilding, J.K. Landeck, and F.G. Calhoun. 1984. Soil Survey of the ICRISAT Sahelian Center, Niger, West Africa. Texas A&M University, TX.
Wilson, S.J. and R.U. Cooke. 1980. Wind erosion. pp. 217–251. *In* M.J. Kirkby and R.P.C. Morgan (eds.) Soil erosion. John Wiley and Sons, London, U.K.

3.4
Effects of Different Rates of Crop Residue Application on Wind Erosion, Soil Properties, and Millet Growth

Ludger Eltrop, Bruce E. Allison, Karlheinz Michels, and M.V.K. Sivakumar

Introduction

Wind erosion damage to soils and plants in the Sahel is well known (e.g. Mainguet and Chemin 1977; Michels 1994; Chapters 2.2 and 3.3). Wind erosion is related to man-made land degradation by deforestation, overcultivation, cultivation of inappropriate land, overgrazing, and to climatic factors. The reduced ground cover is associated with a change in the physical and chemical properties of the soil surface, thus contributing substantially to the lower crop productivity in this region (Raulin 1965; Wendt 1984). In Nigeria, the removal of the topsoil layer by erosion caused a significant decrease in maize yields despite the use of chemical fertilizers (Lal 1985).

Several measures to control wind erosion in the Sahel have been tested, such as ridging, planting of windbreaks, and the application of organic material. While ridging was found to be ineffective (Chapter 3.1), natural savannah windbreaks and planted windbreaks reduced soil flux considerably (Chapters 3.1 and 3.2). However, with natural savannah windbreaks, straw and grain yields of pearl millet did not increase significantly (Chapter 3.1), whereas with planted windbreaks millet yield was increased depending on the windbreak species (Chapter 3.2).

In the Sahel, organic material such as millet residues, branches and twigs are commonly used to reduce wind erosion damage and to improve degraded agricultural areas (Chapter 4.3). Crop residues reduce wind erosion mainly by trapping wind-blown soil particles (Plate 21), by reducing wind drag at the soil surface, and by protecting plants against the impact of saltating soil particles. Therefore, the loss of fertile soil and soil degradation may be effectively reduced. Crop residues as a surface mulch may not only replace the missing

B. Buerkert, B. E. Allison and M. von Oppen (eds.), Wind Erosion in Niger, 139–151.

vegetation cover, but may also provide nutrients for the growing crop (Bationo and Mokwunye 1991; Bationo *et al.* 1993; Chapter 3.3) and stimulate crop growth (Hafner *et al.* 1993; Rebafka *et al.* 1994; Chapter 3.3). Additionally, crop residues may decrease the temperature and physical resistance of the soil surface (Chapter 3.3).

However, crop residues are often not available in sufficient quantities. On farmers' fields in Niger, the mean millet residue production averaged less than 1300 kg ha^{-1} (McIntire and Fussel 1989) and at the onset of the next rainy season only 21 to 39% remained on the fields (ICRISAT 1993). In southwestern Niger, the amount of crop residues at the beginning of the rainy season was only 100 to 250 kg ha^{-1} (Chapter 1.3). The loss of crop residues was mainly due to microbial decomposition, termites, and cattle grazing (Bationo *et al.* 1995; Plate 24). Additionally, farmers use crop residues for other purposes, for example for construction (Plate 25) and firewood (Chapter 4.3).

Quantitative data on the effect of crop residues on wind erosion, soil properties, and crop production in the Sahel are scarce. Therefore, the objective of this study was to determine the effects of different quantities of millet crop residues on soil erosion (soil flux), soil chemical and physical properties, and on the establishment and growth of pearl millet (*Pennisetum glaucum* (L.) R. Br.). In this chapter a summary of the papers reported by Michels *et al.* (1995a, 1995b) is presented.

Effects of Crop Residues on Wind Erosion and Soil Properties

Experimental Site and Wind Erosion Measurements

A field experiment was conducted at the ICRISAT Sahelian Center, Sadoré (Niger) during 1991 and 1992. The total rainfall was 603 mm in 1991 and 585 mm in 1992. The long-term average annual rainfall at the experimental site is 545 mm. The experimental field, an Arenosol (Chapter 1.2), was divided into 18 plots with three treatments arranged in a spatial cross-over design. Crop residues were applied at rates of 500 kg ha^{-1} (comparable to the amount reasonably available in millet-based farming systems) and 2000 kg ha^{-1}. Plots without crop residues served as a control. The treatments were replicated six times. The size of each plot was 19 by 45 m. Prior to the sowing of millet, the millet crop residues were manually placed on the soil surface.

The extent of wind erosion was measured by determining the amount of wind-blown soil at 0.1 m height above ground level with BSNE soil samplers (Fryrear 1986; Plate 12). The soil samplers were installed in the middle of each plot. The wind speed was measured at 2 m height with A100R cup anemometer (Vector Instruments, North Wales) at the ICRISAT weather station, 1 km west of the field.

Soil Flux

In both years, twelve convective storms were counted in the rainy season between May and the beginning of August. The maximum wind speed was 8.2 m s^{-1} in 1991 and 19.6 m s^{-1} in 1992, with a corresponding soil flux on the control plot of 169 kg m^{-2} sampler opening (1991) and 302 kg m^{-2} (1992). Due to a strong sandstorm on 1 June 1992, the crop residues were partially blown away and were subsequently replaced. Compared with the control plots, the total soil flux during the rainy season in 1991 and in 1992 was significantly less on the plots with 2000 kg ha^{-1} of crop residues, whereas on the plots with 500 kg ha^{-1} no significant reduction was observed (Fig. 3.4–1). On the plots with 2000 kg ha^{-1} of crop residues, the decrease in soil flux was 46% in 1991 and 49% in 1992. This supports similar results of Buerkert *et al.* with crop residues and a plastic mulch (Chapter 3.3).

During the dry seasons, major 'Harmattan' storms occurred in December 1991 and January 1992 as well as in March and April 1993. The total soil flux during the 'Harmattan' seasons was distinctly lower than during the rainy seasons, and was significantly reduced in the two crop residue treatments compared to the control (Fig. 3.4–1). Considerable amounts of soil were trapped between the millet stalks.

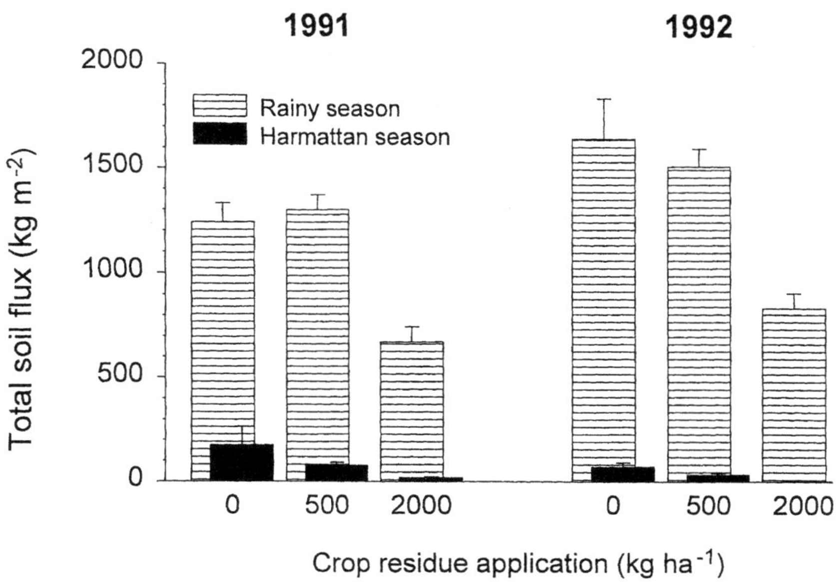

Fig. 3.4–1. Total soil flux at 0.1 m height above ground level during the rainy season and the dry 'Harmattan' season in 1991 and 1992 with applications of 500 and 2000 kg ha^{-1} of crop residues and without a crop residue application (control). Vertical bars indicate mean standard error (SE).

The comparison of the soil flux for the two crop residue treatments indicates that during storms with high wind speeds, crop residues may decrease the soil flux only if applied in amounts exceeding a certain threshold level. Below that level, crop residues had no significant effect on the amount of wind-blown soil. According to Fryrear (1985), at least 10% of the soil must be covered with material to effectively reduce the loss of soil. In this study, this was only the case when 2000 kg ha^{-1} of crop residues were applied (Michels 1994). Only this high amount of crop residues appears to be sufficient to significantly decrease the soil flux during the rainy season. However, the effect of different amounts of crop residues on soil flux may vary with different types of soils and weather conditions. Therefore, the quantity of crop residues needed to effectively control wind erosion in the Sahel, should be evaluated more precisely if practical recommendations are to be given.

The effectiveness of crop residues in controlling wind erosion is not only determined by the amount but also by the mode of placement. Farmers in Niger have different placement practices. Millet heads may be harvested and the straw left on the field until the onset of the rainy season (Plate 5) or the straw may be cut and removed or placed on the ground leaving standing stubble. Alternatively, millet stems including the roots may be pulled out of the soil, left on the field, and burned before the rainy season to obtain a clean field (Plate 26). Crop residues lying flat on the ground were the least effective in controlling wind erosion compared to any other orientation (Siddoway *et al.* 1965). Although not quantified for the Sahel, it is assumed that standing millet straw and stubble decreases soil flux more than flat crop residues, due to the increased surface roughness. Plant density also plays an important role in wind erosion control (Lyles and Allison 1975). In Niger, the density of millet pockets is relatively low (0.3 to 1.5 m^{-2}; Plate 6) and the effects of standing millet straw on wind erosion have yet to be evaluated.

Chemical Soil Properties

After two years of crop residue applications of 2000 kg ha^{-1}, the pH and the concentrations of organic carbon (C), exchangeable calcium (Ca) and magnesium (Mg), as well as the total cation exchange capacity (CEC) significantly increased at the 0 to 0.01 m soil depth compared with the control (Table 3.4–1). The concentrations of exchangeable potassium (K) also tended to be higher. In the plots with 500 kg ha^{-1} of crop residues, soil chemical properties were less affected. Increased concentrations of K, Ca, and Mg and increased pH on plots with 2000 kg ha^{-1} of crop residues were also reported by Buerkert *et al.* (Chapter 3.3) at 0 to 0.1 m soil depth and by Geiger *et al.* (1992) after five years of application at 0 to 0.2 m soil depth. In the study of Geiger *et al.* (1992), the concentration of exchangeable phosphorus (P) was also increased, but in contrast to our study, the concentration of organic C remained unchanged (Geiger *et al.* 1992; Chapter 3.3). The increase in

Table 3.4–1. Chemical properties of soil samples at 0 to 0.01 m soil depth after two years of crop residue applications and of wind-blown soil material captured at 0.1 and 0.35 m height above ground level during the 1992 rainy season

	$pH_{(KCl)}$	$C_{organic}$	$P_{(Bray)}$	Cation exchange capacity (CEC) ($cmol_c\ kg^{-1}$)						
		($g\ kg^{-1}$)	($mg\ kg^{-1}$)	H^+	Al^{3+}	Na^+	K^+	Ca^{2+}	Mg^{2+}	Total
Crop residues										
0 $kg\ ha^{-1}$	4.69	1.16	14.40	0.02	0.04	0.11	0.04	0.46	0.18	0.85
500 $kg\ ha^{-1}$	4.74	1.22	14.12	0.02	0.03	0.11	0.05	0.53	0.17	0.92
2000 $kg\ ha^{-1}$	4.99	1.62	14.14	0.02	0.01	0.11	0.07	0.59	0.22	1.02
ANOVA ($P > F$)	0.05	0.033	0.984	0.153	0.129	0.095	0.117	0.010	0.005	0.004
Wind-blown soil material										
0.10 m height	5.14	1.51	27.8	0.04	0.02	0.12	0.04	0.74	0.19	1.15
0.35 m height	n.d.	2.03	25.4	n.d.	n.d.	0.14	0.04	0.90	0.20	>1.31

n.d. = not determined

organic C in the plots is of particular importance since organic C improves both the water-holding capacity and the CEC, particularly in sandy soils. High amounts of crop residues may improve soil fertility, in part, through the decomposition of millet stalks, by trapping wind-blown soil and dust, and by preventing trapped soil being redistributed by the wind (Chapter 3.3).

Zobeck and Fryrear (1986) also reported increased concentrations of organic C and a higher CEC in wind-blown soil collected at different heights above ground level. The increase in Ca and P concentrations and in CEC in wind-blown soil in our study, are likely due to wind-blown manure or wind-blown fertilizer particles which contain considerable amounts of Ca (Chapter 3.3). Dust that contains more material from faraway had much higher concentrations of mineral elements, particularly of K, Ca, and Mg than wind-blown soil. This resulted in a twelvefold increase in CEC (Chapter 2.1).

Soil Particle Size Distribution

The particle size distribution at 0 to 0.01 m soil depth and in wind-blown soil was determined in November 1991 and November 1992. The sand fractions (2000 to 63 μm) were determined by sieving, the silt and clay fractions (< 63 μm) by sedimentation.

In the control plots, a high proportion of soil particles were in the sand and silt fractions: 11% of the soil particles were of 2000 to 630 μm size, 57% were of 630 to 200 μm size, 27% were of 200 to 63 μm size, and 5% smaller than 63 μm. In 1992, after two years of crop residue application, there was a distinct difference in particle size distribution of the surface soil layer

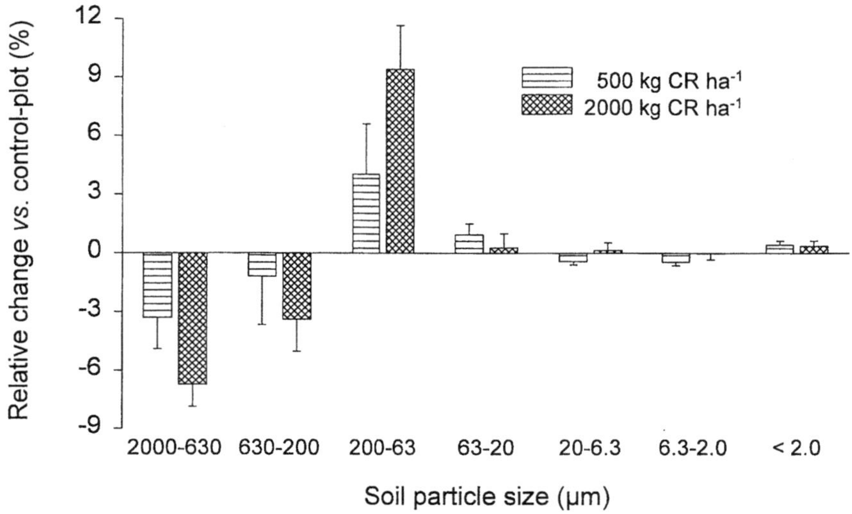

Fig. 3.4–2. Relative change in soil particle size distribution between plots without crop residue application (control) and plots with 500 and 2000 kg ha^{-1} of crop residues. Vertical bars indicate mean standard error (SE).

between crop residue and control plots (Fig. 3.4–2). The fraction of coarse sand (2000 to 200 μm) was significantly reduced and the fraction of fine sand (200 to 63 μm) was significantly increased in the crop residue compared with the control plots. In contrast, the proportion of the fraction smaller than 63 μm was not affected by the crop residue application. The increase in fine sand particles is likely due to the reduced soil erosion in these plots and the entrapment of fine sand particles during sandstorms, as indicated by the particle size distribution of wind-blown soil shown in Chapter 2.2. With increasing height from 0.1 to 0.75 m above ground level, the amount of larger size particles (2000 to 200 μm) decreased from 46 to about 23%. The proportion of medium size particles (200 to 20 μm) increased from 48% at 0.1 m height to 66% at 0.75 m height and the amount of soil particles smaller than 2 μm increased slightly from 3.8% at 0.1 m height to 8.8% at 0.75 m height. In dust deposits collected on fallow land during mid April to mid July, 84% of the particles were smaller than 50 μm (Drees *et al.* 1993). Thus, it appears that dust deposits have distinctly different chemical and physical properties than wind-blown soil distributed during a sandstorm.

Crop Residue Effects on the Establishment and Growth of Pearl Millet

Pearl millet *cv.* CIVT was manually sown in pockets at 10 000 pockets ha^{-1} on 27 May in 1991 and in 1992. Three weeks after sowing, plant density

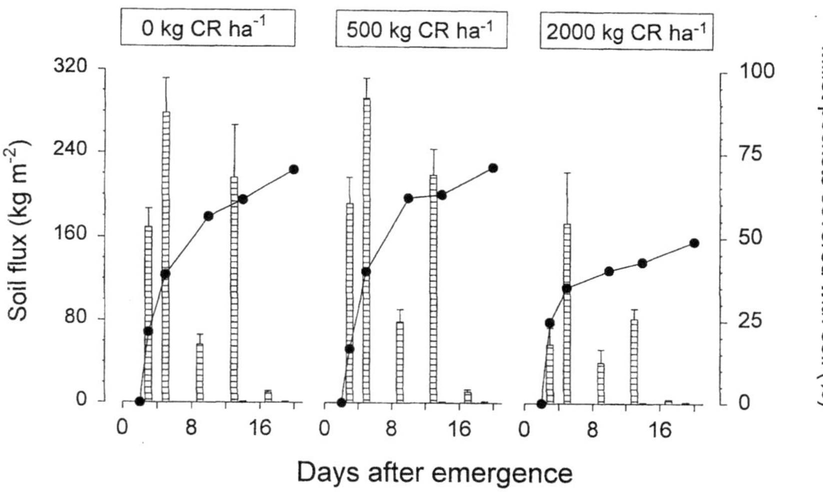

Fig. 3.4–3. Soil flux at 0.1 m height above ground level (hatched bars) and the coverage of millet pockets with soil (•) in control plots without crop residues and in plots with 500 and 2000 kg ha^{-1} of crop residues in the rainy season 1991. Vertical bars indicate mean standard error (SE).

was thinned to three plants per pocket. Nitrogen (N) was applied at a rate of 30 kg ha^{-1} during tillering and at a rate of 15 kg ha^{-1} during shooting. Phosphorus was applied before sowing as single superphosphate at a rate of 18.5 kg P ha^{-1}. The establishment of millet was determined by recording the percentage of millet pockets that were partially or totally covered with soil after each sandstorm. Pockets were considered 'not covered', when at least five seedlings per pocket emerged after the sandstorm. The plants were harvested at physiological maturity.

Coverage of Millet Pockets with Soil and Growth of Pearl Millet

Millet emerged two days after sowing. Heavy sandstorms with high amounts of wind-blown soil occurred three and five days after emergence (Fig. 3.4–3). Even though soil flux was significantly lower on plots with 2000 kg ha^{-1} of crop residues, compared to the bare control or plots with 500 kg ha^{-1} of crop residues, approximately 40% of millet pockets were covered with soil (Fig. 3.4–3). However, during the next sandstorms 9 and 13 days after emergence, a significantly lower percentage of pockets was covered with wind-blown soil on the plots with 2000 kg ha^{-1} of crop residues compared to the other treatments. This supports the conclusions of Eltrop *et al.* (Chapter 2.2) that it is not the extent of the sand flux *per se*, but the time of occurrence during millet establishment that primarily determines the degree of wind erosion damage. Visual observations indicated that mechanical damage (abrasion of leaves) was of less importance and that the extent of soil coverage also depended on plant size. On plots with 2000 kg ha^{-1} of crop residues, the recovery of the more vigorous plants appeared to be better and the proportion of re-emerged plants was higher than on the control plots or the plots with 500 kg ha^{-1} of crop residues. These results support those of Buerkert *et al.* (Chapter 3.3) who found that 2000 kg ha^{-1} of crop residues increased millet dry matter at thinning and booting. The authors concluded that this was attributed to the reduction in nutrient stress, to the physical improvement of the soil by termites, and to the control of water and wind erosion by reducing soil particle movement (Chapter 3.3).

On the plots with 2000 kg ha^{-1} of crop residues, the reduced soil flux may have also decreased the thickness of the soil layer covering the millet plants after a sandstorm. Therefore, plant regrowth after partial or total coverage with soil may have been improved. This assumption is supported by results showing quicker recovery and regrowth of partially covered plants compared to totally covered millet plants (Michels 1994).

Millet Dry Matter Production

Millet dry matter production significantly increased with the application of crop residues in 1991 and 1992 (Fig. 3.4–4). Although in the first observation year, the low rate (500 kg ha^{-1}) failed to improve biomass production, straw and grain yields in 1992 were significantly higher in both crop residue treatments compared to the control. This indicates a cumulative effect of crop residues in 1992. In the plots with 500 kg ha^{-1} of crop residues the straw dry matter increased by 31%, and on the plots with 2000 kg ha^{-1} by 54% compared to the control. Grain yields on the 500 kg ha^{-1} plots increased by 26% and on the plots with 2000 kg ha^{-1} of crop residues by 49%.

In 1993, when wind and water erosion were generally more severe, the application of 2000 kg ha^{-1} of crop residues increased the plant density by 15% and the dry matter production per pocket at maturity threefold (Chapter 3.3). In the present study, plant stands at maturity did not differ significantly among treatments (data not shown). The increased millet dry matter production on crop residue plots was most likely due to increased soil fertility. Kretschmar *et al.* (1991) and Hafner *et al.* (1993) attributed the increase in millet growth with crop residue application, in part to P mobilization and enhanced phytohormone production by soil microorganisms. Since millet contains substantial amounts of P (Rebafka 1993), and the fertility of the acid sandy soils in Niger is mainly determined by the content of organic matter and the concentration of P (Bationo and Mokwunye 1991), small amounts of additionally available P may affect crop yields significantly. Geiger *et al.* (1992) related the increased mineral nutrient concentration in the soil after a crop residue application to increased decomposition rates caused by termites and microorganisms. However, in the present study, the P concentration of

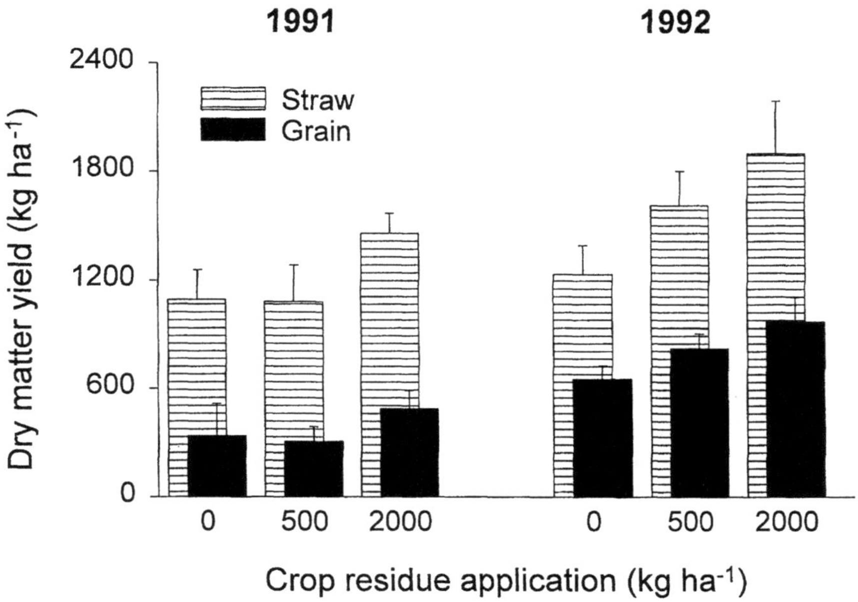

Fig. 3.4–4. Straw and grain dry matter yields of pearl millet at maturity in control plots and in plots with 500 and 2000 kg ha^{-1} of crop residues in 1991 and 1992. Vertical bars indicate mean standard error (SE).

the soil in the crop residue plots was not increased (Table 3.4–1). This may have been due to rapid P cycling between the soil and the plants or to the short duration of the crop residue application, which did not allow subsequent P accumulation in the soil.

The application of crop residues may also significantly affect the water status of the soil. Directly after a rain event, the evaporation rate from a residue covered soil is lower than from a bare soil. This is due to reduced wind speed and an insulating effect (reduced radiation) at the soil surface. Later, the soil evaporation rate may be similar or even higher than on a bare soil because the water remains closer to the soil surface (Unger *et al.* 1991). Bilbro and Fryrear (1991) reported that in the semi-arid Great Plains (USA), the soil water content increased after the application of mulch. However, little corresponding research has been conducted in the Sahel. In the dry season, soil water contents in Niger were not affected by a crop residue mulch (Payne *et al.* 1990). Decreased soil resistance after a crop residue application (Chapter 3.3) may likely increase water infiltration. In a long-term experiment, crop residue applications affected neither soil water-holding capacity nor soil water use by the plants (Bationo *et al.* 1993). Furthermore, crop productivity at the study site is limited more often by low soil fertility than by low soil water content (Bley 1990; Fechter 1993). Thus, in the Sahel the enhancement of millet growth by crop residue applications is likely primarily caused by decreased wind erosion damage and increased soil fertility, and secondly to improved water status of the soil.

In the Sahel, crop residues are often placed only on selected degraded spots of the field (Chapter 4.3; Plate 23) due to the lack of sufficient material. This procedure may increase the millet yield similar to a broad application of 2000 kg ha^{-1} of crop residues, but only on the selected spots. Therefore, this method is not appropriate for reducing wind erosion effectively on a larger scale. The lack of sufficient crop residues appears to be the major constraint for using crop residues for wind erosion control in the Sahel. Despite the application of mineral fertilizers in the present experiment, the production of millet straw was less than 2000 kg ha^{-1}, the amount necessary for an effective wind erosion control. In an on-farm survey in Niger, McIntire and Fussel (1989) found millet straw yields without fertilizer application to be in the range of 470 to 2400 kg ha^{-1}. Average millet straw yields in the northern Sahelian zone of Niger (350 to 425 mm of rainfall) were 1750 kg ha^{-1} and in the southern Sahelian zone (600 mm of rainfall) 2940 kg ha^{-1} (ICRISAT 1992). Livestock grazing, decomposition by termites, and microorganisms as well as household needs reduce this amount to about 100 to 200 kg ha^{-1} (Chapter 1.3). Farmers in the semi-arid regions of Niger may therefore produce sufficient amounts of straw for an efficient wind erosion control only, if external inputs increase the crop production in the Sahel.

Conclusions

The results show that on the sandy soils of the southern Sahelian zone, crop residues may effectively reduce wind erosion and increase millet dry matter production. However, only the application of about 2000 kg $ha^{-1}$of crop residues was found to be effective. At present, it is not realistic to expect this amount to be produced in the Sahel without the external input of fertilizers. Crop residues are also used for a variety of other purposes such as animal feed, building material, firewood, or for selling. Additionally, the burning of crop residues is a common practice in Niger. In order to develop practical recommendations for subsistence farmers in Niger, the economic situation of the farmers must be taken into account. It is also necessary to evaluate, at least in the short run, which combination of crop residue uses will bring optimal returns for the farm operation (Chapter 4.2).

Abstract

In the Sahel region of West Africa, crop residues and other organic materials are commonly used by farmers to rehabilitate degraded areas and to reduce wind erosion. During the 1991 and 1992 growing seasons, a field trial was conducted at the ICRISAT Sahelian Center, Sadoré

(Niger), to investigate the effect of crop residues applied at rates of 500 and 2000 kg ha^{-1} on the extent of wind erosion, on the physical and chemical soil properties, and on the establishment and growth of pearl millet (*Pennisetum glaucum* (L.) R. Br.). Compared to the control plots without crop residues, the total soil flux during the rainy season was significantly reduced on the plots with 2000 kg ha^{-1} of crop residues, but was not significantly different on the plots with 500 kg ha^{-1}. After two years of crop residue application, the proportion of large size soil particles (2000 to 200 μm) was significantly lower at the 0 to 0.01 m soil depth and the proportion of medium size soil particles (200 to 20 μm) was correspondingly higher in both crop residue treatments. In the plots with 2000 kg ha^{-1} of crop residues, the concentration of organic carbon and the cation exchange capacity at the 0 to 0.01 m soil depth were significantly higher than in the control plots. The number of millet seedlings established was significantly higher on the plots with 2000 kg ha^{-1} of crop residues than on the control plots or on the plots with 500 kg ha^{-1} of crop residues. At maturity, the dry matter yield of millet straw and grain was significantly higher in the two crop residue treatments than in the control. The results indicate that a considerable reduction in soil flux and in wind erosion damage is only achieved when high amounts of crop residues are applied. However, in both observation years the dry matter yield of millet straw was less than 2000 kg ha^{-1} on all plots and was, thus, not sufficient to provide the required amount of crop residues for an efficient wind erosion control.

Résumé

Effets de l'utilisation de résidus de récoltes en quantités différentes sur l'érosion éolienne, les caractéristiques du sol et la croissance du petit mil

Dans la région sahélienne d'Afrique de l'Ouest les résidus de récoltes et d'autres matières organiques sont utilisés par les paysans pour réhabiliter les espaces dégradés et pour réduire l'érosion éolienne. Durant les saisons de culture 1991 et 1992, une expérience au champ a été menée au Centre Sahélien ICRISAT, Sadoré (Niger), afin d'étudier les effets de l'utilisation de résidus de récoltes en quantités de 500 et 2000 kg ha^{-1} sur l'étendue de l'érosion éolienne, les caractéristiques physiques et chimiques du sol et sur l'implantation et la croissance du petit mil (*Pennisetum glaucum* (L.) R. Br.). En comparaison des parcelles-témoins nues sans résidus de récoltes, le flux total du sol pendant la saison des pluies a nettement diminué sur les parcelles couvertes de 2000 kg ha^{-1} de résidus de récoltes mais la différence n'a pas été significative sur les parcelles recouvertes de 500 kg ha^{-1}. Après deux années d'application de résidus de récoltes, une quantité nettement moindre de sables grossiers (2000 à 200 μm) et une quantité proportionnellement plus importante de sables moyens (200 à 20 μm) ont été mesurées à des profondeurs de 0 à 0,01 m sur les parcelles concernées par l'application des résidus de récoltes. Sur les parcelles couvertes de 2000 kg ha^{-1} de résidus de récoltes, le taux de carbone organique et la capacité d'échanges cationiques étaient nettement plus élevés à ces mêmes profondeurs que sur les parcelles-témoins. L'implantation de semis de petit mil était nettement plus importante sur les parcelles recouvertes de 2000 kg ha^{-1} de résidus de récoltes que sur les parcelles-témoins ou sur les parcelles recouvertes de 500 kg ha^{-1} de ces résidus. A maturité, la production finale de matière sèche de paille et de grains de petit mil était bien supérieure sur les parcelles à résidus de récoltes que sur les parcelles-témoins. Les résultats indiquent que les effets néfastes du flux de sol et de l'érosion éolienne sur le petit mil peuvent être considérablement réduits, mais seulement si le champ est recouvert de très grandes quantités de résidus de récoltes. Cependant, durant les deux années d'observation, la production de matière sèche de paille de mil était inférieure à 2000 kg ha^{-1} sur toutes les parcelles et ne suffisait pas, par conséquent, à fournir les quantités nécessaires de résidus de récoltes afin de maîtriser l'érosion éolienne.

References

Bationo, A. and A.U. Mokwunye. 1991. Role of manures and crop residue in alleviating soil fertility constraints to crop production: With special reference to the Sahelian and Sudanian zones of West Africa. *Fertilizer Research* 29: 117–125.

Bationo, A., B.C. Christianson, and M.C. Klaij. 1993. The effect of crop residue and fertilizer use on pearl millet yields in Niger. *Fertilizer Research* 34: 251–258.

Bationo, A., A. Buerkert, M.P. Sedogo, B.C. Christianson, and A.U. Mokwunye. 1995. A critical review of crop residue use as soil amendment in the West African semi-arid Tropics. pp. 305–322. *In* J.M. Powell, S. Fernández-Rivera, T.O. Williams, and C. Renard (eds.) Livestock and sustainable nutrient cycling in mixed farming systems of sub-Saharan Africa. II. Technical papers. Proceedings International Conference, Addis Ababa, Ethiopia, 22–26 Nov. 1993, ILCA (International Livestock Centre for Africa), Addis Ababa, Ethiopia.

Bilbro, J.D. and D.W. Fryrear. 1991. Pearl millet versus gin trash mulches for increasing soil water and cotton yields in a semi-arid region. *Journal of Soil and Water Conservation* 46(1): 66–69.

Bley, J. 1990. Experimentelle und modellanalytische Studien zum Wasser- und Nährstoffhaushalt von Perlhirse (*Pennisetum americanum* L.) in Südwest-Niger. Diss. Universität Hohenheim, Stuttgart, Germany.

Drees, L.R., A. Manu, and L.P. Wilding. 1993. Characteristics of aeolian dusts in Niger, West Africa. *Geoderma* 59: 213–233.

Fechter, J. 1993. The simulation of pearl millet (*Pennisetum glaucum* L.) growth under the environmental conditions of south-west Niger, West Africa. Hohenheimer Bodenkundliche Hefte 10, Stuttgart, Germany.

Fryrear, D.W. 1985. Soil cover and wind erosion. *Transactions of the ASAE* 28(3): 781-784.

Fryrear, D.W. 1986. A field dust sampler. *Journal of Soil and Water Conservation* 41(2): 117–120.

Geiger, S.C., A. Manu, and A. Bationo. 1992. Changes in a sandy Sahelian soil following crop residue and fertilizer additions. *Soil Science Society of America Journal* 56: 172–177.

Hafner, H., E. George, A. Bationo, and H. Marschner. 1993. Effect of crop residues on root growth and nutrient acquisition of pearl millet in an acid sandy soil in Niger. *Plant and Soil* 150: 117–127.

ICRISAT. 1992. ICRISAT West African Programme, Annual Report 1991. International Crops Research Institute for the Semi-Arid Tropics, Niamey, Niger.

ICRISAT. 1993. ICRISAT West African Programme, Annual Report 1992. International Crops Research Institute for the Semi-Arid Tropics, Niamey, Niger.

Kretschmar, R.M., H. Hafner, A. Bationo, and H. Marschner. 1991. Long and short-term effects of crop residues on aluminium toxicity, phosphorus availability and growth of pearl millet in an acid sandy soil. *Plant and Soil* 136: 215–223.

Lal, R. 1985. Soil erosion and its relation to productivity in tropical soils. pp. 237–247. *In* S.A. El-Swaify, W.C. Moldenhauer, and A. Lo (eds.) Soil erosion and conservation. SCS, Ankeny, IA.

Lyles, L. and B.E. Allison. 1975. Wind erosion: Uniformly spacing nonerodible elements eliminates effects of wind direction variability. *Journal of Soil and Water Conservation* 30(5): 225–226.

Mainguet, M. and M.C. Chemin. 1977. Les marques de l'érosion éolienne dans le Sahel du Niger d'après les images satellites et les photographies aériennes. pp. 139–149. *In* Colloque pédologie télédétection, AISS, Rome, Italy.

McIntire, J. and L.K. Fussell. 1989. On-farm experiments with millet in Niger: Crop establishment, yield loss factors and economic analysis. *Experimental Agriculture* 25: 217–233.

Michels, K. 1994. Wind erosion in the southern Sahelian zone. Diss. Universität Hohenheim, Verlag Ulrich E. Grauer, Stuttgart, Germany.

Michels, K., M.V.K. Sivakumar, and B.E. Allison. 1995a. Wind erosion control using crop residue. I. Effects on soil flux and soil properties. *Field Crops Research* 40(2): 101–110.
Michels, K., M.V.K. Sivakumar, and B.E. Allison. 1995b. Wind erosion control using crop residue. II. Effects on millet establishment and yields. *Field Crops Research* 40(2): 111–118.
Payne, W.A., C.W. Wendt, and R.J. Lascano. 1990. Bare fallowing on sandy fields of Niger, West Africa. *Soil Science Society of America Journal* 54: 1079–1084.
Raulin, H. 1965. Techniques et bases socio-économiques des sociétés rurales nigériennes. Etudes nigériennes 12, IFAN-CNRS, Niamey, Niger.
Rebafka, F.-P. 1993. Deficiency of phosphorus and molybdenum as major growth limiting factors of pearl millet and groundnut on an acid sandy soil in Niger, West Africa. Hohenheimer Bodenkundliche Hefte 9, Stuttgart, Germany.
Rebafka, F.P., A. Hebel, A. Bationo, K. Stahr, and H. Marschner. 1994. Short and long-term effects of crop residues and of phosphorus fertilization on pearl millet yield on an acid sandy soil in Niger, West Africa. *Field Crops Research* 36: 113–124.
Siddoway, F.H., W.S. Chepil, and D.V. Armbrust. 1965. Effect of kind, amount, and placement of residue on wind erosion control. *Transactions of the ASAE* 8(3): 327-331.
Unger, P.W., B.A. Stewart, J.F. Parr, and R.P. Singh. 1991. Crop residue management and tillage methods for conserving soil and water in semi-arid regions. *Soil and Tillage Research* 20: 219–240.
Wendt, C. 1984. Soil management. p. 15. *In* L.T. West, L.P. Wilding, J.K Landeck, and F.G. Calhoun. Soil survey of the ICRISAT Sahelian Center, Niger, West Africa. Soil and Crop Sciences Department/TropSoils, Texas A&M University System, College Station, TX.
Zobeck, T.M. and D.W. Fryrear. 1986. Chemical and physical characteristics of wind-blown sediment II. Chemical characteristics and total soil and nutrient discharge. *Transactions of the ASAE* 29(4): 1037–1041.

PLATES

Plate 1. A strong convective storm during the rainy season transporting large amounts of soil (Sadoré, western Niger). Photograph by A. Buerkert.

Plate 2. Deforested landscape during the dry season (western Niger). Photograph by J.P.A. Lamers.

Plate 3. A village surrounded by millet fields (foreground) and in the background lateritic plateaus and slopes that have been eroded by water (western Niger). Photograph by J.P.A. Lamers.

Plate 4. Natural savannah vegetation with herbaceous and woody plants. The fenced enclosure has equipment for dust collection (Sadoré, western Niger). Photograph by L. Herrmann.

Plate 5. Strong 'Harmattan' winds cover the landscape with dust during the dry season (near Maradi, southeastern Niger). Photograph by A. Buerkert.

Plate 6. An older dune system (40 000 years old) cropped with pearl millet (near Niamey, western Niger). The dune surface structure is created by surface soil creep. Photograph by A. Buerkert.

Plate 7. *Brousse tigrée*: lateritic plateau with alternating strips of bare laterite and strips of shrub vegetation in small depressions (near Sadoré, western Niger). Photograph by A. Buerkert.

Plate 8. Millet parkland system with *Faidherbia albida* trees growing within millet fields (near Fabridji, western Niger) and an approaching sandstorm. Photograph by J.P.A. Lamers.

Plate 9. Millet is a staple crop to the population in Niger. Bundles of harvested millet heads are stored in huts made from millet stalks (bottom) and *Andropogon* grass (roof) (near Bellaré, western Niger). Photograph by A. Buerkert.

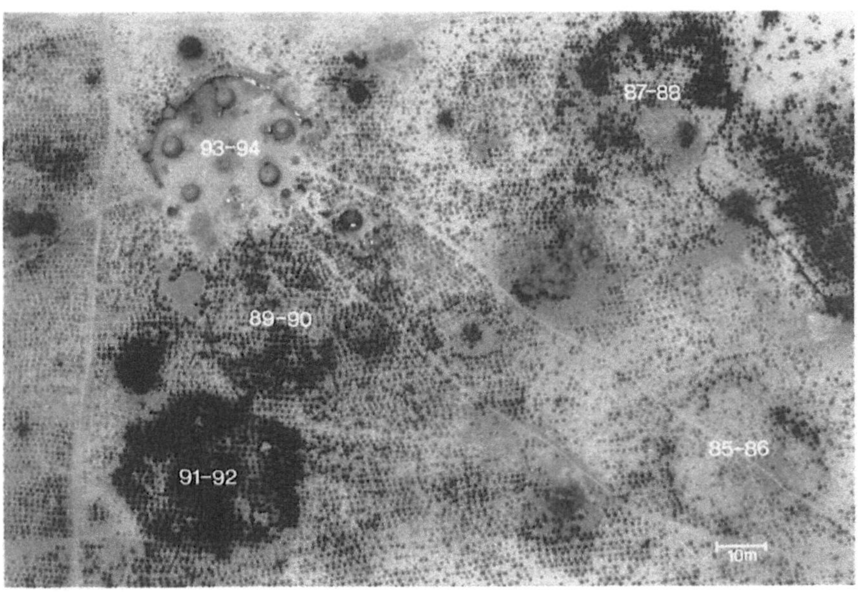

Plate 10. Aerial photograph showing spatial variability in millet growth on a typical field in western Niger. The settlement was moved every second year from site to site as indicated by numbers in the photograph. Photograph by A. Buerkert.

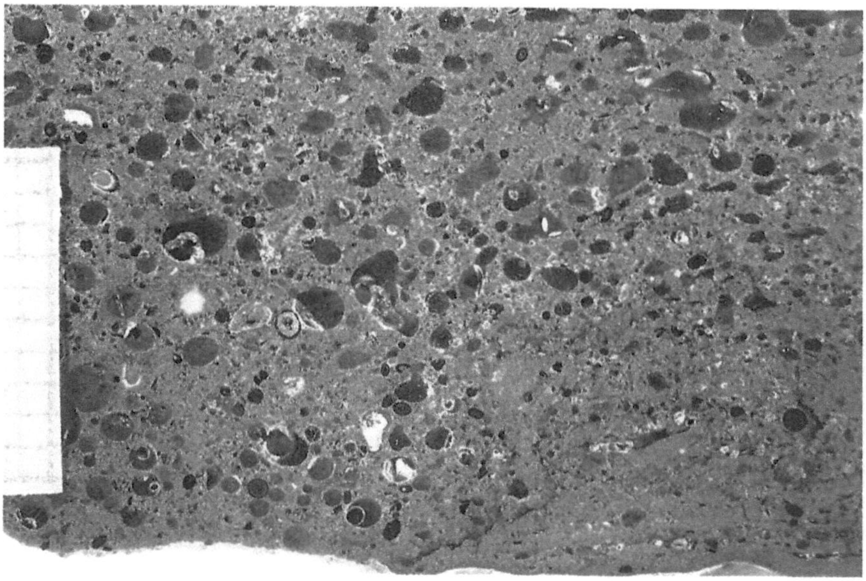

Plate 11. Polished block that shows dust sediment with vesicular pores. Such pores reduce the infiltration of rain water. Photograph by K. Stahr.

Plate 12. Field soil sampler (Big Spring Number Eight, BSNE). Photograph by J.P.A. Lamers.

Plate 13. Soil ridges and young millet plants damaged after a convective storm (Sadoré, western Niger). Photograph by J.P.A. Lamers.

Plate 14. Windbreaks that consist of 4-year-old *Bauhinia rufescens* trees trimmed to 2 m height (windbreak trial at Sadoré, western Niger). Photograph by J.P.A. Lamers.

Plate 15. Windbreaks that consist of 4-year-old *Andropogon gayanus* grass trimmed to 2 m height (windbreak trial at Sadoré, western Niger). Photograph by J.P.A. Lamers.

Plate 16. Windbreak rows that consist of 10-year-old neem (*Azadirachta indica*) trees. One tree row has been recently pruned for cash (Majjia Valley, southern Niger). Photograph by A. Buerkert.

Plate 17. Firewood storage: collected, stacked between grain bins and ready to be sold (near Fillingue, western Niger). Photograph by J.P.A. Lamers.

Plate 18. Small branches of trees and shrubs are transported to the market by donkeys to be sold as firewood (near Niamey, western Niger). Photograph by A. Buerkert.

Plate 19. A neem tree (*Azadirachta indica*) that has been heavily pruned for fodder (see leaves on the ground) and firewood (near Sadoré, western Niger). Photograph by J.P.A. Lamers.

Plate 20. Free-browsing young camels in a millet-parkland-system (Majjia Valley, Niger). The camels prefer green leaves of shrubs and trees to millet stalks. Photograph by J.P.A. Lamers.

Plate 21. Soil deposition between millet crop residues applied at a rate of 2000 kg ha^{-1} during the dry season. The non-protected soil shows development of surface crusts (Sadoré, western Niger). Photograph by A. Buerkert.

Plate 22. Crop residues applied as a mulch on entire fields, is uncommon in western Niger. Photograph by A. Buerkert.

Plate 23. Farmers apply high amounts of millet crop residues and trash on selected areas of their fields. This has resulted in better millet growth compared to areas where no crop residues have been applied (near Sadoré, western Niger). Photograph by A. Buerkert.

Plate 24. Millet crop residues are used for livestock feed after millet harvest. Trees are also used to store cowpea hay (near Sadoré, western Niger). Photograph by A. Buerkert.

Plate 25. Millet stalks are used as building material to construct fences, houses and hangars (western Niger). Photograph by A. Buerkert.

Plate 26. Farmers burn remaining crop residues to obtain a clean field before the next cropping season (near Sadoré, western Niger). Photograph by A. Buerkert.

PART FOUR

Socio-economic Evaluation of Wind Erosion and Control Measures

4.1
Financial Analysis of Seven Planted Windbreaks

John P.A. Lamers and Matthias von Oppen

Introduction

Soil degradation in the West African Sahel is a major constraint to agricultural production and considering the importance of the agricultural sector for the economy of the countries in the region, it threatens economic development. Wind-generated soil erosion is a major concern since it results in the degradation of the predominantly sandy soils. Wind erosion in the Sahel can successfully be reduced by windbreaks that consist of natural bush savannah or planted trees and shrubs (Chapters 3.1 and 3.2). Yet, knowledge about their financial profitability is limited in the Sahel and in other parts of the world (Brandle *et al.* 1988). In addition, most economic or financial analyses do not consider all possible benefits associated with windbreaks (Anderson 1987). Farmers in the Sahel can be expected to adopt and invest in erosion control technologies only if their financial profitability is assured. As shown in Chapter 3.2, in addition to erosion reduction and crop protection, tree windbreak systems produce economically important products such as firewood and animal feed. Farmers in the Sahel may benefit from tree windbreaks in two ways:

(i) Direct benefits of windbreaks
- providing wood for fuel and construction
- supplying feed for livestock production.

(ii) Indirect benefits of windbreaks
- increasing and stabilizing the output of millet grain production
- increasing the output of the by-products of millet production such as straw and herbaceous plants for livestock feed.

In this chapter, seven windbreak species consisting of *Bauhinia rufescens*, *Azadirachta indica*, *Acacia senegal*, *Acacia nilotica*, *Acacia holosericea* and *Faidherbia albida* and the perennial grass *Andropogon gayanus*, described in Chapter 3.2, are analysed. This is done by integrating the *agronomic* (grain, straw, and fodder weeds from the millet fields), *pastoral* (fodder from

B. Buerkert, B. E. Allison and M. von Oppen (eds.), Wind Erosion in Niger, 169–192.

the windbreak canopy), and *silvicultural* (firewood from the windbreaks) uses concurrently. The objective is to estimate the additional net revenues compared to a control without windbreaks and to evaluate the investment profitability for farmers in terms of financial returns, returns to family labor, and the sensitivity to input and output prices. The financial analyses focused on the profitability of windbreaks at the farm level because decisions about an investment in windbreaks are made by farmers.

Methodology of the Financial Analyses

Measures of Profitability

Investments in windbreaks will be economically sound if the values of direct plus indirect benefits, derived from the windbreak-cropping systems, are greater or equal to the cost of capital investments to establish windbreaks plus the production loss when farmland is taken out of production (Brandle *et al.* 1984). Establishment and use of an agroforestry system, such as the windbreak-millet system, involves several years and, therefore, it is necessary to compare costs and revenues over time (Hart 1991). As the cash flows of the examined windbreaks change over the investment period, they are discounted to a common point in time (here the present) which allows for a direct comparison of the costs and revenues. The discount rate is the rate by which future sums of money, spent on inputs or received as outputs, are discounted to determine the present value. It was assumed that the economic minimum lifetime of the windbreaks was 15 years, during which the trees were not destroyed by fire, cutting, drought, or any other disaster (for further assumptions see Appendix 4.1–1).

Three frequently-used financial indicators of returns in the evaluation of agroforestry systems are the present value of incremental net returns (PVINR), the internal rate of return (IRR), and the benefit/cost ratio (BCR). These were also employed in this study to measure profitability of investments in windbreaks. At positive PVINRs, an investment in windbreaks is recommendable to farmers because such windbreaks produce net returns which are greater than the discount rate used. Farmers should refrain from an investment in windbreaks when the PVINR is negative. The IRR calculates a discount rate at which the present values of costs and returns just break even. The method of calculating the IRR has an advantage over the PVINR method because it avoids the difficulty of choosing a discount rate. An investment in windbreaks is recommendable if the computed discount rate is greater than the real rate of interest. The BCR relates the present value of total benefits and total costs and thus, represents the gain per invested monetary unit. An investment is considered worthwhile if the BCR is greater than 1. PVINR estimations provide information about the actual financial volumes involved, while BCR and IRR do not provide such information.

The choice of the discount rate, which reflects the rate of interest or the opportunity cost of capital, is of major importance because the outcome of a financial analysis based on net present values depends on the discount rate used. In 1991, the real rate of interest (adjusted for inflation) in Niger was about 7% (World Bank 1993). In 1991, money could be borrowed from a Nigerien bank at a rate of 11% (Statistisches Bundesamt 1992). Development projects apply interest rates of 9 to 12% for their different credit programs. Other evidence from developing countries showed that the real discount rate varied between 8 and 15% (Gittinger 1982). In the base scenario of our analyses, a discount rate of 10% was used, but to account for the lack of capital which is commonly a limiting factor for the resource poor households in the Sahel (Chapter 1.3), cash flows were also computed with discount rates of 15 and 30%.

Quantification of Benefits and Costs

The incremental revenues (gains or losses) are calculated as the difference in net revenues (total benefits minus costs of millet production) between the windbreaks and the control without windbreaks. The data to determine benefits and costs were monitored in the experiment described in Chapter 3.2 from 1991 to 1993. Detailed benefits and costs associated with the production from the millet fields and the windbreaks are summarized in Appendix 4.1–2. The production from the control was adjusted for the area (11.7%) not occupied by windbreak strips. For the main features of the trees species examined, see Appendix 4.1–3.

Benefits and costs of the base and the sensitivity analyses are valued at constant market prices which are presented in Table 4.1–1. The prices were either monitored during a market study or taken from literature. Since there is no market for browse materials, shadow prices were calculated based on the metabolizable energy (ME) concentration in the feed dry matter (MJ ME kg^{-1} DM) (de Montgolfier-Kouèvi and Le Houérou 1980). To test the sensitivity of the results in the base scenario, the measures of profitability were also calculated while varying the prices for millet grain, firewood, browse material, and tree seedlings. When performing the sensitivity analyses, only one component at the time was altered (Table 4.1–1).

Total establishment costs of the different tree windbreaks consisted of expenses for seedlings (9550 FCFA for 227 trees), material during planting (2400 FCFA), hired labor for planting (7580 FCFA) and maintenance weeding (between 1000 and 1200 FCFA, depending on the tree species) in the first four years, and totalled around 20 700 FCFA ha^{-1} (Appendix 4.1–4).

Table 4.1–1. Commodity prices and the price of tree seedlings used in the base scenario and the sensitivity analysis of the financial analyses of seven windbreak species in southwest Niger

Item[a]	Price base scenario	Prices assumed in sensitivity analyses				Unit
Millet grain	80	20	40	120	160	FCFA kg^{-1}
Millet straw	10	–	–	–	–	FCFA kg^{-1}
Firewood	7.5	0	4	15	22.5	FCFA kg^{-1}
Fodder browses	4.5	0	2	9	13.5	FCFA ME^{-1}[b]
Fodder weeds	16.5	–	–	–	–	FCFA kg^{-1}
Tree seedling	42	0	84	126	–	FCFA per tree

[a] *Source*: grain (OPVN 1992); straw, firewood, and fodder weeds (Lamers 1995); fodder browses estimated according to de Montgolfier-Kouèvi and Le Houérou (1980); tree seedling (Stolwijk 1989)
[b] ME = Metabolizable energy (MJ kg^{-1} DM)

Table 4.1–2. Scenarios examined in the financial analyses of seven planted windbreaks in southwest Niger

	Indirect benefits	Direct benefits	Costs
Scenario	Included millet grain, straw, and edible weeds	Included firewood and fodder	Additional labor due to windbreaks covered by
1	yes	no	family labor
2	yes	yes	family labor
3	yes	no	hired labor
4	yes	yes	hired labor

Scenarios

Four scenarios were examined in the financial analyses (Table 4.1–2). The most likely situation for Nigerien smallholders would be represented by *scenario 2* (Baidu-Forson 1988; McIntire *et al.* 1989). The sensitivity of the results for millet grain prices were evaluated with the assumptions defined in *scenario 1*. The sensitivity analyses for the direct benefits were computed under the assumptions of *scenario 2*. Family labor in *scenarios 1* and *2* was valued at zero opportunity costs. When assuming that the additional labor requirements for pruning and maintenance weeding of the windbreaks were covered by hired labor (*scenarios 3* and *4*), a wage rate of 1000 FCFA day^{-1} was imputed. A working day was assumed to be 8 h.

Farmers in Niger aim at maximizing labor productivity rather than discount rates of capital. An important criterion for adoption is, therefore, the returns to family labor. A decision on investing in windbreaks may strongly depend on these returns, as they represent the best use of the family resources compared to off-farm alternatives or other farm activities. Returns to family labor were calculated as averages over 1991, 1992, and 1993. Both the indirect (*scenario 1*) and the direct plus indirect effects (positive or negative) of the windbreak-millet system (*scenario 2*) were evaluated.

Financial Returns to Windbreaks

Returns to *A. gayanus* and *B. rufescens* windbreaks were lower than the control, as demonstrated in all four scenarios by the negative PVINRs as well as BCRs less than 1 at a discount rate of 10% (Table 4.1–3). In contrast, returns to the *F. albida* windbreak were greater than the control for all four scenarios, as evidenced by positive PVINRs as well as BCRs greater than 1 and IRRs greater than 7%. The IRRs of an investment in *F. albida* varied between 27% (*scenario 3*) and 35% (*scenario 2*).

Returns to *A. indica*, *A. senegal*, *A. holosericea*, and *A. nilotica* windbreaks depended on the specific scenarios. Given only the incremental benefits derived from the millet field under the assumptions of *scenario 1*, returns to *F. albida* and *A. holosericea* windbreaks showed positive PVINRs as well as BCRs greater than 1, and IRRs of 20% (*F. albida*) and 32% (*A. holosericea*). When additional labor was given a monetary value (*scenario 3*), only returns to *F. albida* windbreaks were profitable as indicated by an IRR of 27%.

When the direct plus the indirect effects were computed without valuing additional labor (*scenario 2*), *A. indica*, *A. senegal*, *A. holosericea*, *F. albida*, and *A. nilotica* windbreaks showed positive PVINRs, had BCRs greater than 1, and had IRRs that were more than two to five times the 7% interest rate in Niger (Table 4.1–3). When the additional labor for the direct plus indirect effects were given a monetary value (*scenario 4*), only *A. indica*, *A. holosericea*, and *F. albida* windbreaks had positive PVINRs as well as BCRs greater than 1 and an IRR of 16% (*A. indica*) or greater (*F. albida* and *A. holosericea*).

The indirect effects of windbreaks on grain and straw yields depended on the tree species (Chapter 3.2). Windbreaks such as *A. gayanus* and *B. rufescens* reduced millet grain yields compared to the control without windbreaks. This effect is particularly disliked by farmers whose subsistence is often put to question. Lower yields in the presence of windbreaks also indicate a lower investment capacity of farmers because food purchase normally depletes cash reserves (McIntire *et al.* 1989). On the other hand, higher grain yields

Table 4.1–3. Present value of incremental net return (PVINR) and benefit / cost ratio (BCR) at a discount rate of 10%, as well as the internal rate of return (IRR) of four production / variable cost scenarios for the seven windbreak species, estimated over an assumed lifetime of 15 years for the species in southwest Niger

	Unit	*Andropogon gayanus*	*Bauhinia rufescens*	*Azadirachta indica*	*Acacia senegal*	*Acacia holosericea*	*Faidherbia albida*	*Acacia nilotica*
***Scenario 1:* indirect benefits / family labor**								
PVINR	FCFA ha^{-1}	–49 236	–45 978	–12 995	–36 512	17 320	48 207	–6 918
BCR	–	–2.6	–2.4	0.7	–1.5	3.7	6.5	1.3
IRR	%	–	–	–	–	20	32	4
***Scenario 2:* direct plus indirect benefits / family labor**								
PVINR	FCFA ha^{-1}	–47 373	–31 210	27 610	11 691	61 054	58 189	16 204
BCR	–	–2.4	–1.0	4.6	3.0	8.0	7.4	3.5
IRR	%	–	–	23	16	35	35	19
***Scenario 3:* indirect benefits / hired labor**								
PVINR	FCFA ha^{-1}	–50 496	–73 073	–31 202	–13 929	–30 728	33 718	–80 961
BCR	–	–2.3	–0.7	0.3	–0.1	0.6	2.7	0.2
IRR	%	–	–	–	–	–	27	–
***Scenario 4:* direct plus indirect benefits / hired labor**								
PVINR	FCFA ha^{-1}	–48 634	–58 306	9 403	–91 089	13 006	43 700	–57 838
BCR	–	–2.2	–0.3	1.7	0.3	1.4	3.1	0.4
IRR	%	–	–	16	–	18	30	–

in windbreak-millet cropping systems would increase both the investment capacity of farmers and the potential to satisfy household food demands. In view of the controversial short-term effects of windbreaks on agricultural production, Kaisin (1994) argued that windbreaks are not a general solution to improve the livelihood of Sahelian households. However, he considered only the indirect effects coming from the protected crops.

Although windbreaks may compete with the millet crop and thus be detrimental in biological terms, windbreaks can provide a financial benefit. Dennison (1986) and Madougou *et al.* (1989) economically evaluated 11-year-old neem (*A. indica*) windbreaks in the Majjia Valley in south-central Niger. They concluded that there was an annual benefit of about 5000 FCFA ha^{-1} including millet production, firewood supply, and wooden poles for construction. Wood production was the most beneficial component, but it was difficult for farmers to realize these returns without felling the trees, which are the capital investment themselves. Therefore after eleven years, the benefits from wood were only partially available to farmers. An economic evaluation in northern

Table 4.1–4. Returns to family labor (in FCFA h^{-1}) of indirect (millet grain and straw, and fodder weeds) and indirect plus direct (firewood and fodder) benefits of the seven windbreaks species and the control in southwest Niger

Returns to family labor	Control	*Andropogon gayanus*	*Bauhinia rufescens*	*Azadirachta indica*	*Acacia senegal*	*Acacia holosericea*	*Faidherbia albida*	*Acacia nilotica*
Indirect benefits	274*	248	246	287	287	348	395	305
– millet grain only	218	191	196	226	226	273	304	242
Direct plus indirect benefits	27	250	264	332	332	399	407	334

*Means over 1991, 1992, and 1993

Nigeria showed profitable returns to windbreaks when accounting for their multiple effects (Anderson 1987). A financial analysis of three agroforestry systems (*Tectona grandis*, *Gmelina arborea*, *Eucalyptus calmaldulensis*) as practised by smallholders in Nigeria, demonstrated positive net present values and internal rates of return equal to or greater than 12% when the costs and benefits of the annual crops (yam, maize, cassava, millet, sorghum, beans) and the perennial trees were simultaneously considered (Osemeobo 1989). The present analyses also confirm the importance of the indirect effects for the profitability of windbreaks. The loss in millet production for the *A. gayanus* and *B. rufescens* windbreaks was not compensated for by the direct benefits of material harvested from these windbreaks. However, the large firewood and fodder production of *A. indica* could compensate for the financial loss due to the reduced millet production in the presence of this species. Also, the indirect effects of *A. senegal* and *A. nilotica* windbreaks were counterbalanced by their direct benefits as long as additional labor demands were not covered by hired labor (Table 4.1–3). In order to profit from the direct effects of the windbreaks, labor for pruning the trees is required. Thorny species such as *A. senegal* and *A. nilotica* had the highest labor demands.

Returns to Family Labor

The financial analyses accentuated the role of additional labor requirements for the profitability of the windbreak-millet cropping system. Returns to family labor varied between 246 FCFA h^{-1} (*B. rufescens*) and 395 FCFA h^{-1} (*F. albida*) when only the indirect benefits were considered, and between 250 FCFA h^{-1} (*A. gayanus*) and 407 FCFA h^{-1} (*F. albida*) when the direct plus indirect benefits were considered (Table 4.1–4). The returns to family labor of *A. gayanus* and *B. rufescens* windbreaks were smaller than the control without windbreaks.

In our analyses, returns to family labor were higher than the wage rate for a hired laborer in western Niger villages, which varied between 66 and

138 FCFA h^{-1} depending on the village and the year (McIntire *et al.* 1989). However, comparisons using household surveys need some additional consideration. Millet straw production, for instance, was not financially valued in the village survey and, consequently, returns to family labor were lower. When straw was excluded from the evaluation in our study, the returns to family labor decreased between 20% (control without windbreaks) and 23% (*A. gayanus* and *F. albida*), but it never went below the returns from the village survey.

On the other hand, millet grain yields recorded in the on-station experiment (Chapter 3.2) were much higher than in the villages surveyed (McIntire *et al.* 1989). Taking the grain yield differences into account, the returns to family labor in our study were on the same level as the returns to family labor in western Niger villages. Based on the grain yield in the windbreak experiment, and including the financial value of millet straw in the analysis, the returns to family labor were at least similar to some minimum hourly wages for adult laborers, varying between 123 and 210 FCFA h^{-1} (Statistisches Bundesamt 1992). Thus, the returns to family labor on the different windbreaks were no worse than existing alternatives.

Sensitivity to Input and Output Prices

Sensitivity to Millet Grain Prices

Farmers are interested in having stable profitability measures for changing grain prices. The sensitivity analyses demonstrated that changes in millet grain prices did not alter the ranking of the windbreak species, but only the level of the expected profitability (Table 4.1–5). At grain prices higher than the market price of 80 FCFA kg^{-1}, the IRR and the PVINR for all windbreak species increased.

Sensitivity to Firewood Prices

Because the firewood consumption in Niger is currently greater than the wood production of forest land, tree planting is necessary (Projet Energie II 1991). Firewood already has an economic potential in Niger as evidenced by the existence of firewood markets at urban centers, by specialized merchants who operate the market channels, and by the high number of firewood vendors (one vendor for every 100 urban households) in Niamey (Projet Energie II 1991). Unlike the rural population, urban dwellers have to pay for firewood. When firewood prices were increased from 7.5 to 22.5 FCFA kg^{-1} the returns to the highest producing firewood species, *A. senegal* and *A. indica,* increased (Table 4.1–6). This agrees with findings reported by Dennison (1986) for *A. indica* windbreaks in Niger. Changes in returns to *A. holosericea*, *F. albida,*

Table 4.1–5. Internal rate of return (in %) of three windbreak species in southwest Niger at different millet grain prices

Price (FCFA kg^{-1})	*Acacia holosericea*	*Faidherbia albida*	*Acacia nilotica*
20	10	19	–
40	14	24	1
80*	20	32	4
120	25	38	7
160	30	44	9

*A millet grain price of 80 FCFA kg^{-1} reflects the base situation of *scenario 1*

Table 4.1–6. Internal rates of return (in %) of five windbreak species in southwest Niger at different firewood prices

Price (FCFA kg^{-1})	*Azadirachta indica*	*Acacia senegal*	*Acacia holosericea*	*Faidherbia albida*	*Acacia nilotica*
0	18	6	35	34	18
4	21	12	35	34	19
7.5*	23	16	35	35	19
15	28	23	35	35	20
22.5	32	28	36	36	20

*A firewood price of 7.5 FCFA kg^{-1} reflects the base situation of *scenario 2*

and *A. nilotica* windbreaks were minor due to their limited firewood production (Appendix 4.1–3). Returns to *A. gayanus* (no firewood production) and *B. rufescens* (medium firewood producer) windbreaks remained lower than the control at any price analyzed in the study.

The sensitivity analyses indicate that if the profitability of investments in windbreaks become more dependent on the direct benefits, such as firewood, the choice of the windbreak species becomes more important. Given the predictions of a population growth above 3% for Niger (Statistisches Bundesamt 1992), firewood will become scarce. Thus, it can be expected that prices for firewood will increase in the future. Firewood prices in the urban centers are largely determined by the transportation costs. Firewood collectors in the proximity of urban centers already gain an additional income from transporting (Plate 18) and trading firewood (Projet Energie II 1991).

Table 4.1–7. Internal rate of return (in %) of five windbreak species in southwest Niger at different fodder prices

Price (FCFA ME^{-1a})	*Azadirachta indica*	*Acacia senegal*	*Acacia holosericea*	*Faidherbia albida*	*Acacia nilotica*
0	11	–	21	32	6
2	17	6	28	33	13
4.5[b]	23	16	35	35	19
9	32	27	44	37	27
13.5	38	36	51	39	34

[a] Metabolizable energy (MJ kg^{-1} DM)
[b] A fodder price of 4.5 FCFA ME^{-1} reflects the base situation of *scenario 2*

Sensitivity to Fodder Prices

In contrast to firewood, no market for browse material exists in Niger. The absence of a market indicates that no private property rights can be claimed on browse material (Plates 19 and 20). Trees may be browsed by herds from passing nomads (Niamir 1990). When the shadow price for fodder varied between 0 and 13.5 FCFA ME^{-1}, the IRR changed for all windbreak species (Table 4.1–7). This was particularly true for *A. senegal*, which was the species that contained high amounts of metabolizable energy per kg leaf dry matter (Lamers *et al.* 1994). Both *A. holosericea* and *A. indica* produced higher amounts of leaf dry matter than *A. senegal*, but the nutritive value of the former species was lower (Lamers *et al.* 1994). Whereas *A. holosericea* contained the lowest content of metabolizable energy of all species, *A. indica* is often rejected by animals as the leaves contain azadirachtine and nimbidine. These chemicals are valued as an insecticide, but probably reduce the palatability (Höfs 1992).

The sensitivity of fodder prices for the *F. albida* windbreaks was lower than for other windbreak species. The direct benefits of this species can be expected during another period of the year since *F. albida* provides few leaves in April, the period of pruning. As shown in Chapter 4.3, both farmers and their wives not only agreed on the importance of trees for livestock production in general, but also that *F. albida* was the species with the best nutritive value. *F. albida* was also the most preferred species for livestock feed by farmers in the Sudano-Sahelian zone of Burkina Faso (März 1993).

Given the importance of livestock production in the Sahel and the decline in natural pasture, leaves from trees and shrubs are likely to increase in value as a livestock feed. Leaves and fruits from *F. albida*, *Z. mauritiana*, *C. aculeatum*, and *B. rufescens*, which were used in feeding trials as a nitrogen source, improved the rumen function of sheep and cattle (Le Houérou 1980;

Table 4.1–8. Internal rates of return (in %) of five windbreak species in southwest Niger at different prices for tree seedlings

Price (FCFA per tree)	*Azadirachta indica*	*Acacia senegal*	*Acacia holosericea*	*Faidherbia albida*	*Acacia nilotica*
0	35	25	50	49	29
42*	23	16	35	35	19
84	17	11	27	27	13
126	13	8	22	22	10

*A seedling price of 42 FCFA per seedling reflects the base situation of *scenario 2*

Höfs 1992). Both studies reported an increase in the digestion of a basal diet of millet and sorghum crop residues. Thus, leaves may have an additional value, particularly in the dry season which is characterized by low quality feed. In our study, this additional value could not be determined because no feeding trials using different proportions of millet straw and browse leaves were done to determine the optimal combination. Thus, the financial value of the leaves based on the metabolizable energy content alone, represents the lowest value of its potential nutritional and economic value. If livestock feed increases in value, it may, consequently, influence more investment decisions by farmers.

Although it is often stated that trees have multiple uses, the present financial analyses indicates that using trees for some particular purpose, such as livestock feed, might reduce the gains from other produce, such as firewood. The IRRs from *A. holosericea* windbreaks varied from 35 to 36%, the IRRs from *F. albida* from 34 to 36%, and the IRRs from *A. nilotica* windbreaks from 18 to 20% (Table 4.1–6). Thus, they were not sensitive to changes in firewood prices compared to the IRRs from *A. senegal* and *A. indica* windbreaks. However, by changing fodder prices, the IRR from *A. nilotica* windbreaks varied from 6 to 34% and the IRR from *A. holosericea* from 21 to 50%, both being much more sensitive to changes in fodder prices than *A. indica* windbreaks (Table 4.1–7). The returns to *A. senegal* windbreaks were sensitive to changes in both firewood and fodder prices.

Sensitivity to Establishment Costs

A variation in tree seedling prices changed the amplitude of the profitability of the windbreaks but not their ranking (Table 4.1–8). The seedling price used to estimate the establishment costs of windbreaks accounted for approximately 45% of the total establishment costs. If tree seedlings were provided at no cost, the IRRs varied between 25 and 50%. If the establishment costs of windbreaks were reduced, for example by lowering seedling prices, the investment risks for farmers decreased. Consequently, positive returns can be realized in a much shorter period of time. However, there is a period of time before any return can be expected. A recent study indicated that due to browsing by livestock, seedling survival of unprotected *A. gayanus* and *Z. mauritiana* plantings was only 45 and 30%, respectively (ICRISAT 1992). It is, therefore, likely that there will be establishment costs for the farmers to absorb even if trees are supplied at no cost. If establishment costs increase to a level higher than those used in the present analysis, the profitability of *A. senegal* and *A. nilotica* windbreaks will become marginal compared with *F. albida* and *A. holosericea*. This is an additional reason for selecting the species carefully.

Direct seeding can reduce establishment costs and improve the profitability of windbreaks. This technique does not require the high investment costs of seedling nurseries which need equipment and labor for seedling production, expensive logistical support during seedling distribution and the knowledge of skilled labor during planting (Sluiter 1989; Stolwijk 1989). Little information is available on direct seeding in the Sahel. Generally though, it is not possible with all species, normally has a lower rate of germination, demands higher quantities of seed, and is more dependent on favorable conditions for germination after planting (Ritchie 1988). Therefore, it is highly recommended that research focus on the establishment and growth constraints of directly seeded windbreak trees.

Sensitivity to Discount Rates

Discount rates of more than 10% reduce the profitability of *A. holosericea* windbreaks when labor is given a monetary value (Table 4.1–9). At discount rates of 30% or higher, only an investment in *F. albida* windbreaks remained a profitable enterprise with positive PVINRs under *scenarios 1*, *2*, and *4*, although the PVINRs were close to zero. At a discount rate of 10%, only the PVINR of *F. albida* and *A. holosericea* were positive in our analyses. This indicated that an investment would be justified at millet grain yield increments of more than 5%. At a discount rate of 30%, only *F. albida* provided a positive return, which indicated that millet yield increments of at least 17% were necessary.

Other well-known benefits of the windbreak species studied could not be taken into account in the current analyses, as the information to quantify these specific benefits was not available. These benefits include pods of *F. albida* for livestock feed, gum arabic of *A. senegal*, bark of *A. nilotica* for tanning leather, timber of *A. indica*, straw of *A. gayanus*, and medicinal properties of all species. Several parts of trees and shrubs are sold at markets not only in Niger but also in other parts of West Africa (Guinko and Pasgo 1992). Despite the multiple uses of trees and shrubs, little information is documented on the economic value of the woody and non-woody parts of trees (Guinko and

Table 4.1–9. Present value of incremental net return (PVINR) and benefit / cost ratio (BCR) at a discount rate of 15 and 30%, as well as the internal rate of return (IRR) of four production / variable cost scenarios for the seven windbreak species, estimated over an assumed lifetime of 15 years, in southwest Niger

	Unit	*Andropogon gayanus*	*Bauhinia rufescens*	*Azadirachta indica*	*Acacia senegal*	*Acacia holosericea*	*Faidherbia albida*	*Acacia nilotica*
***Scenario 1:* indirect benefits / family labor**								
PVINR-15%	FCFA ha^{-1}	–39 593	–37 205	–14 298	–30 944	1 031	27 826	–10 263
BCR-15%	–	–1.5	–1.4	0.4	–0.9	2.1	3.8	0.7
PVINR-30%	FCFA ha^{-1}	–25 250	–24 115	–14 886	–22 030	–6 368	1 509	–13 523
BCR-30%	–	–0.6	–0.5	0.1	–0.4	0.7	1.3	0.2
***Scenario 2:* direct plus indirect benefits / family labor**								
PVINR-15%	FCFA ha^{-1}	–38 336	–27 263	13 016	1 595	36 235	34 557	5 408
BCR-15%	–	–1.4	–0.6	2.6	1.7	4.5	4.3	2.0
PVINR-30%	FCFA ha^{-1}	–24 778	–20 481	4 935	–10 013	4 412	3 985	–7 645
BCR-30%	–	–0.5	–0.3	0.8	0.5	1.5	1.5	0.7
***Scenario 3:* indirect benefits / hired labor**								
PVINR-15%	FCFA ha^{-1}	–40 443	–55 478	–26 545	–100 292	–25 696	18 057	–60 349
BCR-15%	–	–1.4	–0.6	0.2	–0.1	0.6	2.1	0.1
PVINR-30%	FCFA ha^{-1}	–25 564	–30 838	–19 347	–47 595	–18 426	–2 083	–32 169
BCR-30%	–	–0.6	–0.3	0.1	–0.1	0.4	1.0	0.1
***Scenario 4:* direct plus indirect benefits / hired labor**								
PVINR-15%	FCFA ha^{-1}	–39 187	–45 536	797	–67 753	3 757	24 788	–44 677
BCR-15%	–	–1.4	–0.3	1.3	0.3	1.2	2.4	0.4
PVINR-30%	FCFA ha^{-1}	–25 101	–27 204	–9 396	–35 578	–7 647	393	–26 291
BCR-30%	–	–0.5	–0.2	0.6	0.2	0.8	1.2	0.3

Pasgo 1992). Returns to windbreak systems would increase if these features were included.

Long-term Aspects of Windbreaks in the Sahel

The World Bank estimated a 7% interest rate as the opportunity cost of capital in Niger (World Bank 1993). However, capital growth through interest alone is not a realistic investment opportunity to Nigerien farmers. The traditional method of capital accumulation is to invest in livestock production. The few studies available demonstrated returns to capital investments by livestock of 20.5 and 23.5% in the northern regions of the Ivory Coast (Schleich *et al.* 1987). These returns were higher than the computed IRRs for most windbreak species in this study, except for *F. albida*. Compared to the high returns to capital from investments in livestock, investments in tree windbreaks appear, at present, less profitable to farmers. However, the limited carrying capacity of the natural resources will probably restrict the possibility of capital accumulation through livestock and will be less recommendable as an investment possibility in the future. Livestock production at the expense of staple production is not feasible in the long run. The current state of soil degradation in the Sahel requires immediate intervention to stop the process and restore the fragile ecosystems. Investments in livestock should, therefore, go hand in hand with investments in soil conservation and restoration practices to reduce the risk of total livestock loss.

A reduction of wind erosion will improve soil conditions in the long run and, consequently, increase the financial value of the protected land. Yet, there is very little information available on long-term effects of soil erosion control on crop production. In Mali, Zimbabwe, and Nigeria, soil loss was expressed in monetary terms by converting the amount of nitrogen, phosphorus, and potassium lost through soil erosion into the equivalent amount of fertilizer needed to replace the loss (Hailu and Runge-Metzger 1992). Ehui *et al.* (1990) believed soil erosion was responsible for financial losses due to the reduction of crop yields. This would require a long-term relationship between yield and cumulative soil loss (Ehui *et al.* 1990). The present financial evaluation did not consider the salvage value of the protected land because the chemical soil properties of the windbreak and the control plots did not differ after five years, except in the vicinity of the windbreaks (Chapter 3.2).

The calculation of the salvage value based on the loss of nutrients is suitable in cases where soil nutrients are not exported from the land. However, this method is more difficult to apply if the protected soil is cultivated and nutrients are exported from the field in harvested products. Yield increases are anticipated through the protection of millet by windbreaks, but this will be associated with an increase in the export of nutrients. If these nutrients are not replaced, the depletion of the soil will occur at a faster rate than without windbreaks. The litter from the trees and shrubs could provide some organic matter and nutrients to compensate for the loss of nutrients. A mulch with leaves from *A. indica* and *Albizia lebbeck* increased sorghum grain yields in Burkina Faso (Tilander 1993). The mulch from *A. indica* at 1.2, 2.5, and 3.7 t leaf dry matter (LDM) ha^{-1} and from *A. lebbeck* at 0.9, 1.8, and 2.7 t LDM ha^{-1} equaled 25, 50, and 75 kg nitrogen ha^{-1}, respectively. In Niger, such production levels were only recorded at large tree densities of 1943 trees ha^{-1} (Lamers *et al.* 1994). Thus, a mulch of litter seems a profitable activity, but the labor demands required for collecting and distributing the mulch were not considered in the Burkina Faso study (Tilander 1993). However, they proved to be decisive in the profitability of windbreaks in the present study.

A direct soil cover, using a mulch of millet crop residues (Chapters 3.3 and 3.4) or branches from trees and shrubs (Chase and Boudouresque 1987), was efficient in reducing wind and water erosion on Sahelian soils. A mulch of dead material might therefore be more effective than living strips of perennial vegetation consisting of herbaceous plants, shrubs, or trees (Vlaar 1992). The living material competes with the crop for water and nutrients, both limiting factors in the northern Sahelian zone (Kessler and Breman 1991). Consequently, to obtain a better impact from soil protection techniques, any anti-erosion technique should be combined with an input of nutrients, either by mineral or organic fertilizers or a combination of both, particularly on the poor soils predominating in the Sahel.

Conclusions and Final Remarks

The question whether an investment in windbreaks is economically sound can be answered only with some reservations since long-term yield results were not available. The financial returns to windbreaks based on observations during a 3-year period were found to be highly dependent on yield gains of the protected millet. Millet grain and straw yields in the presence of *F. albida* and *A. holosericea* windbreaks were higher than with the control without windbreaks, whereas yields with *A. gayanus* and *B. rufescens* windbreaks were lower than the control. Hence, *F. albida* and *A. holosericea* windbreaks had the highest returns when only the indirect effects from the millet field were considered. The profitability depended also on the products from the windbreaks. *A. senegal* and *A. indica* windbreaks had a good potential for investments in firewood production, whereas *A. senegal* and *A. nilotica* windbreaks had the best potential as feed for animal production. Access to reliable markets for these commodities would be necessary to fully exploit the benefits. Consequently, the presence of a market for firewood and feed may be important for the adoption of a windbreak-millet system. A general implementation of windbreaks without considering the species could result in an increase, a decrease, or no change in farmers' income. Protecting the sandy soils in the Sahel and implementing erosion reducing techniques are best accomplished when mineral fertilizers are used to compensate nutrient export due to higher grain yields (Lamers 1995). This approach better ensures that returns to investments in erosion control measures are maximized.

The allocation of the additional labor involved in the maintenance and pruning of the windbreaks determines the financial returns to the windbreaks as well. Because farmers maximize labor productivity rather than the discount rate of capital, decisions of farmers to invest in windbreaks may strongly depend on the returns of tree windbreaks to family labor. Returns of windbreaks to family labor were no less than existing alternatives, although the returns to family labor for the *A. gayanus* and *B. rufescens* windbreaks were lower than the control.

The additional advantages of wildlife habitat, shade for resting, and other benefits often credited to windbreaks, as well as indirect costs from uncontrolled wind erosion, were not included in the present analyses, but they should also be considered. Russel *et al.* (1990) stated that the indirect costs of erosion to society are significant and, consequently, strongly recommended they be included in economic analyses. However, the valuation procedures for such benefits and costs are rather tedious, as are the methods of analysis. The capital budget approaches used, based on net present values of cash flows, may be inappropriate when considering complex issues important to society as a whole. Net present value methods hypothesize that future benefits gradually reduce in worth, which may be an inappropriate assumption. A major consideration will therefore be the choice of a proper discount rate. On the other hand, profitability for a farmer in the Sahel is primarily concerned with the direct and indirect effects on the value of his farm. Land-use decisions by farmers in the Sahel, who rely on land for a living, differ from opinions of people in more developed countries, who might be more concerned with degradation and erosion aspects and the improvement of the rural economy as a whole (Shaxson 1985). There is no question that prevention of erosion is always the best solution, and the current state of soil degradation demands a solution to minimize the erosion process and rebuild a sustainable farming system. Even though some erosion control techniques are currently not economically feasible, this should not have the highest priority in decision-making.

Abstract

In this chapter, the profitability at farm level of the seven windbreak species described in Chapter 3.2 is compared. The analysis was based on three measurements: the present value of incremental net return (PVINR), the benefit/cost ratio (BCR), and the internal rate of return (IRR). The first two measures of profitability assume an economic lifetime of 15 years for the windbreaks. The comparison evaluates the indirect (grain, straw, and fodder weeds from the pearl millet fields) and the direct (firewood and fodder from the windbreaks) benefits of windbreaks averaged over the years 1991 to 1993. The yield level of the protected millet was an important factor influencing the profitability of the windbreaks. Increases from windbreaks in millet grain yield of 6% or more gave positive PVINRs. Windbreaks consisting of *Andropogon gayanus* and *Bauhinia rufescens* had poor returns with negative PVINRs as well as BCRs less than 1 at discount rates of 10% and above. In contrast, returns to *Faidherbia albida* windbreaks were always high enough to generate positive PVINRs as well as BCRs greater than 1 and an IRR greater than 7%. The IRRs of an investment in this type of windbreak varied between 27% (only direct benefits) and 35% (direct plus indirect benefits). Returns to *Acacia indica, Acacia senegal, Acacia holosericea,* and *Acacia nilotica* windbreaks depended on the benefits and the discount rate. Although windbreaks may compete with the crop and thus be detrimental in biological terms, they can provide a financial benefit. *A. senegal* and *A. indica* had the best potential for investments in firewood projects, whereas *A. senegal* and *A. nilotica* had the best potential as browse material for animal production. Returns to family labor varied between 246 FCFA h^{-1} (*B. rufescens*) and 395 FCFA h^{-1} (*F. albida*) when considering only the indirect benefits, compared to a return of about 274 FCFA h^{-1} for the control without windbreaks. When considering the direct plus the indirect benefits of the windbreaks, the returns to family labor varied between 250 FCFA h^{-1} (*A. gayanus*) and 407 FCFA h^{-1} (*F. albida*). Windbreaks can be a financially sound enterprise, but the agro-climatic as well as the socio-economic constraints need to be considered when selecting the species.

Résumé

Analyse financière de sept brise-vent plantés

Ce chapitre compare la rentabilité à l'échelon de l'exploitation de sept espèces de brise-vent, décrites dans le chapitre 3.2., à partir de trois indices : la valeur actualisée des revenus nets futurs (PVINR), le ratio bénéfice-coût (BCR) et le taux de rentabilité interne (IRR). Les deux premiers indices de rentabilité supposent une durée de vie économique des brise-vent de 15 ans. La comparaison prend en compte les bénéfices indirects (grain, paille et herbes fourragères des champs de petit mil) et les bénéfices directs (bois de feu et fourrage des brise-vent) calculés sur les moyennes des années 1991 à 1993. Le niveau de rendement du petit mil protégé était un facteur important influençant la rentabilité des brise-vent. Des augmentations de 6% ou plus dans la production de grains de mil ont donné des PVINRs positives. Les résultats des brise-vent de *Andropogon gayanus* et *Bauhinia rufescens* ont été décevants avec des PVINRs négatives et des BCRs inférieurs à 1 à un taux d'escompte de 10% et plus. En revanche, les résultats des brise-vent de *Faidherbia albida* ont toujours été suffisants pour générer des PVINRs positives, des BCRs supérieurs à 1 et un IRR supérieur à 7%. Les IRRs d'un investissement dans ce type de brise-vent varient de 27% (bénéfices directs uniquement) à 35% (bénéfices directs et indirects). Les résultats des brise-vent de *Acacia indica, Acacia senegal, Acacia holosericea,* et *Acacia nilotica* ont dépendu des bénéfices et des taux d'escompte. Bien que certains brise-vent soient en compétition avec la culture (et donc à son détriment en termes biologiques) ils peuvent générer un bénéfice financier. *A. senegal* et *A. indica* ont fourni le meilleur potentiel dans les investissements des projets de bois de feu, alors que *A. senegal* et *A. nilotica* ont manifesté un meilleur potentiel comme pâturage dans la production animale. La rémunération du travail familial a varié de 246 FCFA h^{-1} (*B. rufescens*) à 395 FCFA h^{-1} (*F. albida*) si l'on ne prenait en compte que les revenus indirects en comparaison de la rémunération qui s'élève à environ 274 FCFA h^{-1} sur les parcelles-témoins sans brise-vent. Si l'on prenait en compte les revenus directs et indirects des brise-vent, la rémunération du fravail familial a varié de 250 FCFA h^{-1} (*A. gayanus*) à 407 FCFA h^{-1} (*F. albida*). Les brise-vent peuvent se révéler une entreprise financièrement saine, mais nécessitent un choix judicieux des espèces si l'on tient compte des contraintes tant agro-climatiques que socio-économiques.

References

Anderson, D. 1987. The economics of afforestation: A case study in Africa. Published for the World Bank, Washington D.C., Johns Hopkins University Press, Baltimore, MD.

Baidu-Forson, J. 1988. Characteristics of farm households and economy in western Niger villages: Some evidence from baseline surveys. Resource Management Program, Economics Group, Progress Report 87, International Crops Research Institute for the Semi-Arid Tropics, Niamey, Niger.

Brandle, J.R., B.R. Johnson, and D.D. Dearmont. 1984. Windbreak economics: The case of winter wheat production in eastern Nebraska. *Journal of Soil and Water Conservation* 39(5): 339–343.

Brandle, J.R., D.L. Hintz, and J.W. Sturrock. 1988. Windbreak technology. Proceedings International Symposium on Windbreak Technology, Lincoln, Nebraska, 23–27 June 1986, Elsevier Science Publishers B.V., Amsterdam, The Netherlands. Also published as Volumes 22/23 of *Agriculture, Ecosystems and Environment*, 1988.

Chase, R. and E. Boudouresque. 1987. Methods to stimulate plant regrowth on bare Sahelian forest soils in the region of Niamey, Niger. *Agriculture, Ecosystems and Environment* 18: 211–221.

Dennison, S. 1986. Local benefits of windbreak plantings in the West African Sahel. Paper presented at the International Symposium on Windbreak Technology, Lincoln, Nebraska, 23–27 June 1986 (unpublished).

Ehui, S.K., B.T. Kang, and D.S.C. Spencer. 1990. Economic analysis of soil erosion effect in alley cropping, no-till and bush fallow systems in south-western Nigeria. *Agroforestry Systems* 34: 349–368.

Gittinger, J.P. 1982. Economic analysis of agricultural projects. EDI Series in Economic Development, Johns Hopkins University Press, Baltimore, MD.

Guinko, S. and L.P. Pasgo. 1992. Récolte et commercialisation des produits non ligneux des essences forestières locales dans le département de Zitenga, province d'Oubritenga, Burkina Faso. pp. 17–25. *In* R. Wittig and S. Guinko (eds.) Etudes sur la flore et la végétation du Burkina Faso et des pays avoisinants. Vol. I. Verlag Natur & Wissenschaft, Solingen, Germany.

Hailu, Z. and A. Runge-Metzger. 1992. Sustainability of land-use systems: The potential of indigenous measures for the maintenance of soil productivity in sub-Sahara African agriculture. Tropical Agroecology 7, Margraf Verlag, Weikersheim, Germany.

Hart, C. 1991. Practical forestry for the agent and surveyor. 3rd edition. Alan Sutton Publishing Inc., Wolfeboro Falls, USA.

Höfs, S. 1992. Evaluation of various indigenous supplements in millet leaf based diets for sheep in the Sahel. Diss. Technische Universität Berlin, Institut für Tierproduktion, Fachgebiet für Internationale Agrarentwicklung (D38) 239, Germany.

ICRISAT. 1992. ICRISAT West African Programme, Annual Report 1991. International Crops Research Institute for the Semi-Arid Tropics, Niamey, Niger.

Kaisin, E.F.P. 1994. An assessment of the potential of windbreaks for rainfed agricultural development in the African Sahel. pp. 214–217. *In* Proceedings 4th International Symposium on Windbreaks and Agroforestry, Viborg, Denmark, 26–30 July 1993, Hedeselskabet, Denmark.

Kessler, J.J. and H. Breman. 1991. The potential of agroforestry to increase primary production in the Sahelian and Sudanian zones of West Africa. *Agroforestry Systems* 13: 41–62.

Lamers, J.P.A. 1995. An assessment of wind erosion control techniques in Niger. Financial and economic analyses of windbreaks and millet crop residues. Diss. Universität Hohenheim, Verlag Ulrich E. Grauer, Stuttgart, Germany.

Lamers, J.P.A., K. Michels, and R.J. Vandenbeldt. 1994. Trees and windbreaks in the Sahel: Establishment, growth, nutritive and calorific values. *Agroforestry Systems* 26(3): 171–184.

Le Houérou, H.N. 1980. Agroforestry techniques for the conservation and improvement of soil fertility in arid and semi-arid zones. pp. 433–435. *In* H.N. Le Houérou (ed.) Browse in Africa – The current state of knowledge. Proceedings Symposium, ILCA, Addis Ababa, Ethiopia, 8–12 April 1980, International Livestock Centre for Africa, Addis Ababa, Ethiopia.

Madougou, Z., A. Solheim, D. Steinberg, and R.M. Rochette. 1989. Les brise-vent de la Maggia et de Maiguizaoua, Tahoua-Maradi/Niger. pp. 197–220. *In* R.M. Rochette (ed.) Le Sahel en lutte contre la désertification: Leçons d'expériences. CILSS, Comité Inter-Etats de Lutte contre la Sécheresse au Sahel, Margraf Verlag, Weikersheim, Germany.

März, U. 1993. The determination of mixing ratios in afforestation assortments: Satisfying farmers' needs for multi-purpose forests. *Tropenlandwirt* 94: 175–183.

McIntire, J., J. Hopkins, J.P.A. Lamers, and L.K. Fussell. 1989. The millet system of western Niger. I. Crop production. ILCA, Addis Ababa, Ethiopia (unpublished).

Montgolfier-Kouèvi de, C. and H.N. Le Houérou. 1980. Study on the economic viability of browse plantations in Africa. pp. 449–464. *In* H.N. Le Houérou (ed.) Browse in Africa – The current state of knowledge. Proceedings Symposium, ILCA, Addis Ababa, Ethiopia, 8–12 April 1980, International Livestock Centre for Africa, Addis Ababa, Ethiopia.

Niamir, M. 1990. Herders' decision-making in natural resources management in arid and semi-arid Africa. Community Forestry Note No. 4, Food and Agricultural Organization of the United Nations, Rome, Italy.

OPVN. 1992. Marchés des céréales et du niébé au Niger, 1990/1992. Office des Produits Vivriers du Niger: Système d'Information sur les Marchés (S.I.M.), Niamey, Niger.

Osemeobo, G.J. 1989. An impact and performance evaluation of smallholder participation in tree planting, Nigeria. *Agroforestry Systems* 29: 117–135.

Projet Energie II – Energie Domestique. 1991 Volet Offre. Schema Directeur d'Approvisionnement en Bois Energie de Niamey. Groupement SEED-CTFT, République du Niger, Ministère de l'Hydraulique et de l'Environnement, Niamey, Niger.

Ritchie, K.A. 1988. Shelterbelt plantings in semi-arid areas. *Agriculture, Ecosystems and Environment* 22/23: 425–440.

Russell, I.W., A.M.N. Izac, and R.A. Cramb. 1990. Towards an evaluation of the off-side costs of soil erosion in Queensland. Agricultural Economics Discussion Paper, Department of Agriculture, University of Queensland, Australia.

Schleich, K., H. Bassewitz Graf von, A. Bahm, P. Filius, and G. Weinschenk. 1987. Entwicklung der Tierproduktion in Savannegebieten Westafrikas: Studien aus den Norden der Elfenbeinküste. Deutsche Gesellschaft für Technische Zusammenarbeit (GTZ) (ed.), Sonderpublikation (SP) No. 196, Eschborn, Germany.

Shaxson, T.F. 1985. Erosion, economics, subsistence and psychology. pp. 667–673. *In* S.A. El-Swaify, W.C. Moldenhauer, and A. Lo (eds.) Soil erosion and conservation. Soil Conservation Service, Aukeny, IA.

Sluiter, H. 1989. Des normes forestiers. Direction Départemental de l'Environnement, Département de Tahoua, Niger.

Statistisches Bundesamt. 1992. Länderbericht Niger 1992. Verlag J.B. Metzler und C.E. Poeschel, Wiesbaden, Germany.

Stolwijk, C. 1989. Le prix de revient d'un plant produit dans les différents types des pépinières dans le cercle de Baroueli-Mali. Rapport fin d'étude. Projet Bois de Village. Direction Régionale des Eaux et Forêts, Segou-Mali, et Hogeschool Utrecht, Enseignement Supérieur, The Netherlands.

Tilander, Y. 1993. Effects of mulching with *Azadirachta indica* and *Albizia lebbeck* leaves on the yield of sorghum under semi-arid conditions in Burkina Faso. *Agroforestry Systems* 24: 277–293.

Vlaar, J.C.J. 1992. Les techniques de conservation des eaux et des sols dans les pays du Sahel. Rapport d'une étude effectuée dans le cadre de la collaboration entre le Comité

Interafricain d'Etudes Hydrauliques (CIEH), Ouagadougou, Burkina Faso, et l'Université Agronomique Wageningen (UAW), Wageningen, The Netherlands.
World Bank. 1993. World Development Report 1993. Oxford University Press, Washington D.C.

Appendix 4.1–1. Assumptions

According to the windbreak constellation in the experiment, 1 ha of land was divided between windbreak strips (1170 m^2 or 11.7%) and land designated to millet production (8830 m^2 or 88.3%). The grain and straw production of the control without windbreaks was thus increased by 11.7%. The direct and indirect benefits of the integrated windbreak-millet system are indeed computed on a per ha basis, but represent the addition of both surfaces. For the financial analyses, the following assumptions were made:

(i) For some species (*F. albida*, *A. indica*) the assumed lifetime of 15 years is underestimated, but for other species (*A. gayanus*, *A. holosericea*) 15 years is over-estimated.

(ii) Windbreaks became effective from the fourth year onward after they reached a height of 2 m.

(iii) Tree windbreaks were trimmed to 2 m each year between the end of the dry season and the onset of the rainy season: a period of relative pasture shortage and high firewood demand. Trimming of windbreaks occurred with the principal objective of maintaining, in all circumstances, a functional windbreak at the onset of the rainy season. There was, thus, no attempt at maximizing the direct benefits.

(iv) The tree density of all windbreaks was 1943 trees ha^{-1}. Windbreak strips consisted of 227 trees per 1170 m^2 (11.7% of the surface) and could theoretically protect a 20 m wide field strip, up to a total surface of 8830 m^2 (88.3% of the surface).

Appendix 4.1–2. Estimated benefits and costs of seven windbreak millet systems from 1991 to 1993 in southwest Niger

	Unit	Control	*Andropogon gayanus*	*Bauhinia rufescens*	*Azadirachta indica*	*Acacia senegal*	*Acacia holosericea*	*Faidherbia albida*	*Acacia nilotica*
Benefits and costs 1991									
Fodder weeds	FCFA ha^{-1}	705	1 181	1 271	1 845	1 432	1 432	1 432	1 432
Millet	FCFA ha^{-1}	6 376	7 460	5 960	7 390	6 937	6 937	6 937	6 937
Firewood	FCFA ha^{-1}	n.a.	n.a.	n.a.	n.a.	n.a.	n.a.	n.a.	n.a.
Fodder	FCFA ha^{-1}	n.a.	n.a.	n.a.	n.a.	n.a.	n.a.	n.a.	n.a.
Labor lopping	h per tree	n.a.	n.a.	n.a.	n.a.	n.a.	n.a.	n.a.	n.a.
Labor weeding	h ha^{-1}	n.a.	n.a.	n.a.	n.a.	n.a.	n.a.	n.a.	n.a.
Labor millet	h ha^{-1}	147	138	148	153	148	148	148	148
Benefits and costs 1992									
Fodder weeds	FCFA ha^{-1}	722	1 607	2 132	2 903	2 214	2 214	2 214	2 214
Millet	FCFA ha^{-1}	84 541	59 880	66 860	78 890	57 770	96 990	102 150	76 200
Firewood	FCFA ha^{-1}	n.a.	102	852	2 046	2 216	511	170	85
Fodder	FCFA ha^{-1}	n.a.	315	1 774	4 331	9 538	7 623	1 994	7 882
Labor lopping	h per tree	n.a.	0.01	0.31	0.18	0.84	0.47	0.11	0.47
Labor weeding	h ha^{-1}	n.a.	66.3	60.6	57.1	60.6	50.3	58.3	52.6
Labor millet	h ha^{-1}	129	112	118	123	118	118	118	118

Appendix 4.1–2. Continued

	Unit	Control	*Andropogon gayanus*	*Bauhinia rufescens*	*Azadirachta indica*	*Acacia senegal*	*Acacia holosericea*	*Faidherbia albida*	*Acacia nilotica*
Benefits and costs 1993									
Fodder weeds	FCFA ha^{-1}	689	754	410	787	651	651	651	651
Millet	FCFA ha^{-1}	12 605	18 220	14 130	18 280	27 730	19 540	33 170	26 400
Firewood	FCFA ha^{-1}	n.a.	102	1 705	4 423	5 626	170	852	682
Fodder	FCFA ha^{-1}	n.a.	320	2 344	7 568	4 328	11 457	1 485	1 704
Labor lopping	h per tree	n.a.	0.01	0.22	0.19	0.79	0.29	0.12	0.40
Labor weeding	h ha^{-1}	n.a.	66.3	60.6	57.1	60.6	50.3	58.3	52.6
Labor millet	h ha^{-1}	119	143	136	143	140	140	140	140
Mean benefits and mean costs for 1994–2003									
Fodder weeds	FCFA ha^{-1}	705	1 181	1 271	1 845	1 432	1 432	651	1 432
Millet	FCFA ha^{-1}	34 507	28 520	28 983	34 853	30 812	41 156	26 400	36 512
Firewood	FCFA ha^{-1}	n.a.	102	2 059	3 239	3 921	341	682	384
Fodder	FCFA ha^{-1}	n.a.	318	1 271	5 950	6 933	9 540	1 704	4 793
Labor lopping	h per tree	n.a.	0.01	0.27	0.19	0.82	0.37	0.12	0.43
Labor weeding	h ha^{-1}	n.a.	66.3	60.6	57.1	60.6	50.3	58.3	52.6
Labor millet	h ha^{-1}	132	131	134	140	129	129	129	129

(*Source*: Lamers 1995, summarized)
n.a. = not available

Appendix 4.1–3. Main characteristics of the tree species used as windbreaks in southwest Niger

Characteristic	*Andropogon gayanus*	*Bauhinia rufescens*	*Azadirachta indica*	*Acacia senegal*	*Acacia holosericea*	*Faidherbia albida*	*Acacia nilotica*
Mortality rate[a]	-	-	-	-	++	-	-
Growth rate[a]	++	+	++	+	++	-	+
Wood production[a]	none	+	++	++	+	+	+
Firewood value index[a]	-	+	+	-	n.a.	-	++
Leaf production[a]	- -	+	++	+	++	+	+
Crude protein (CP)[a]	-	+	++	++	+	+	+
Metabolizable energy (ME)[a]	+	+	++	+	-	++	++
CP/ME ratio[a]	-	+	+	++	+	+	+
Labor for lopping	- -	+	+	++	+	+	++
Labor for maintenance weeding of windbreaks	++	++	+	++	-	+	-
Short-term effects on millet dry matter production[b]	-	-	+	+	+	++	-

(*Sources*: [a]Lamers 1995, summarized; [b]Eltrop *et al.* Chapter 3.2)

- - very low; - low; + medium; ++ high

n.a. = not available

Appendix 4.1–4. Input prices and total inputs used for windbreak planting to protect one hectare in 1988 in Niger

Item	Input price	Unit	Total input[b]	Unit
Tree seedling	42[a]	FCFA per tree	227	trees
15–15–15 fertilizer	65	FCFA kg^{-1}	0.86	kg
Furadan® (pesticide)	65	FCFA kg^{-1}	1.13	kg
Water	5	FCFA l^{-1}	454	FCFA
Labor for planting	1000	FCFA per day[d]	7.56[c]	day
Labor for weeding	1000	FCFA per day[d]	1–1.2	day

[a] *Source*: Stolwijk (1989, p. 39)
[b] Input necessary to plant windbreak strips of 1170 m^2 (11.7% of the surface) according to assumption 4
[c] Based on the observation that one laborer could handle 30 seedling per day (Sluiter 1989, p. 6)
[d] A working day of 8 h

4.2 An Economic Evaluation of the Use of Millet Crop Residues from a Farming Systems Perspective

Michael Bruentrup, John P.A. Lamers, Franz Heidhues, and Matthias von Oppen

Introduction

In the technology research projects at the ICRISAT Sahelian Center, Niger, the effects of different methods of controlling wind erosion in the Sahel were recently investigated. Windbreaks consisting of different tree species and grasses, and a surface application of millet crop residues were found to be effective in reducing wind erosion. However, the effects on grain and straw yields depended very much on the windbreak design and the amount of crop residues applied (Chapters 3.1 to 3.4). In Chapter 4.1, it was concluded that the financial returns to windbreaks were dominated primarily by the grain and straw yields of the protected millet crop, and secondly by the products from windbreaks such as firewood and feed. Thus, if innovations for controlling wind erosion in the Sahel such as windbreaks or mulching with crop residues are to be accepted by farmers, they should not reduce family income.

In traditional Nigerien farming systems, the crop residues needed to reduce the effects of wind erosion have alternative uses, for example as fodder, building material, to gain cash, or are burned to obtain a clean field (Chapter 4.3). Knowledge about the economics of the uses of crop residue in the Sahel is rare. This study aims to fill this gap by comparing the use of crop residues for wind erosion control and alternative uses from a farming systems perspective.

Apart from the effect on crop production, an important consideration for farmers in assessing new technologies is the labor requirements involved. Mellor and Johnson (1984) stated that labor availability is often the key constraining factor in African farming systems. For any evaluation of innovations to be relevant, it should consider the profitability of both land and

B. Buerkert, B. E. Allison and M. von Oppen (eds.), Wind Erosion in Niger, 193–214.

labor. As indicated in Chapter 3.3, the nutritional effects of a crop residue application of 2000 kg ha^{-1} may be more important to millet growth than the mechanical effects. However, the following analysis only considered a combination of mulch effects, including protection against wind erosion damage and improving chemical and physical soil properties. Furthermore, our analysis was restricted to a short-term evaluation of crop residues because long-term data on dry matter production and different soil management techniques were not available.

The analyses were performed in two steps. First, the gross margins were calculated to compare the economic profitability of the alternative uses, which were expressed per unit land area, per unit total labor, and per unit of labor for the two weeding operations. Second, a linear programming (LP) model was developed to simultaneously evaluate the profitability of these alternative uses. The LP model searches for that combination of activities that gives the highest possible total gross margin with given resource endowments and activities. The general mathematical formulation of a linear programming model is as follows:

$$\textit{Maximise } Z = \sum_{j=1}^{n} c_j x_j$$

subject to the linear constraints :

$$\sum_{i=1}^{m} \sum_{j=1}^{n} a_i x_j \leq b_i \quad \text{with } i = 1 \text{ to } m$$

$$x_j > 0 \quad \text{with } j = 1 \text{ to } n$$

Z = the value of the objective function, which will be maximized
x = the activities, here the alternative uses of crop residues
c = the coefficients of the objective function, here the costs of the activities
a = the matrix coefficients specifying the resource requirements of the activities x
b = the resource availabilities or restrictions.

The prices and revenues were presented at the prevailing rates prior to the devaluation of the FCFA in January 1994, i.e. 1 French Franc equals 50 FCFA and 1 US$ about 270 FCFA.

Table 4.2–1. Gross margins for land and labor at different application rates of crop residues

	Unit	*Zero mulch*[b]	*Low mulch*	*High mulch*
Input				
Labor for weeding (1st + 2nd)	h ha^{-1}	52	55	64
Total labor	h ha^{-1}	179	165	192
Additional crop residues[a]	kg ha^{-1}	0	0	824
Output				
Millet grain	kg ha^{-1}	496	567	732
Crop residues[a]	kg ha^{-1}	1165	1348	1680
Gross margins with monetary valuation of crop residues				
Per unit land	FCFA ha^{-1}	49 303	50 594	48 837
Per unit total labor	FCFA h^{-1}	275	307	254
Per unit weeding	FCFA h^{-1}	948	920	763
Gross margins without monetary valuation of crop residues				
Per unit land	FCFA ha^{-1}	39 680	45 360	58 560
Per unit total labor	FCFA h^{-1}	221	275	304
Per unit weeding	FCFA h^{-1}	763	825	915

[a] Crop residue balance: if output ×0.7 (loss rate = 0.3) exceeded the necessary crop residue input, surplus crop residue production was counted as output. On the contrary, the additional crop residues to complement 2000 kg were valued as input. Crop residues are valued at 11.8 FCFA per kg

[b] *Zero mulch* = bare control without crop residues; *low mulch* = 500 kg ha^{-1} of crop residues; *high mulch* = 2000 kg ha^{-1} of crop residues

Gross Margins of Mulching with Crop Residues *versus* Their Complete Removal

The first step in the economic evaluation was to compare the gross margins of mulching with crop residues to the complete removal of this material. The remuneration of land and labor was computed based on the input and output values of the different uses of the crop residues (Table 4.2–1). The valuation of these inputs and outputs was based on market prices (Table 4.2–2). Additional labor associated with the different uses of crop residues was considered to be covered by family labor only. Labor remuneration was calculated as the residual income. Labor time (h ha^{-1}) for the *zero mulch* and *high mulch* uses was measured directly from agronomic trials (Chapter 3.4) conducted in 1991 and 1992 on plots of 750 m^2 in size. Labor time for cultivating was nearly the same for *low mulch* as for *zero mulch*. Yet, labor time for the two weeding operations was assumed to increase proportionally to the amount of

Table 4.2–2. Market prices of farm inputs and outputs

Item	Price per unit (FCFA)	Unit	Remarks
Grain[a]	80	kg	annual average
Crop residues[b]	11.8	kg	dry season average
Meat[c]	224	kg live-weight	dry season average
Labor[d]	700	day (= 6 hours)	average agricultural activities
Fertilizer (SSP)[e]	65	kg	

(*Sources*: [a]OPVN (1992); [b]Authors' survey on Niamey market 1992/1993; [c]Ministère des Ressources Animales (1991); [d]Authors' survey; [e]Purchase price of single superphosphate by the ICRISAT Sahelian Center)

crop residues applied. For the *low mulch* use, an additional weeding time of 25% of the difference between *zero mulch* and *high mulch* use was calculated.

Observations showed that the amount of crop residues between harvest and the beginning of the succeeding rainy season was reduced by 30%, thus reducing the availability of crop residues for all uses evaluated. This reduction was caused primarily by termite activities and wind blow. For the *zero mulch* use, it was assumed that all crop residues were sold. For the *low mulch* use however, only surplus crop residues were sold. The agronomic experiments revealed that *high mulch* use did not reproduce the needed amount of millet crop residues (Chapters 3.3 and 3.4) and, consequently, some amounts had to be purchased or imported from other fields on the farm. For purchase and sale, 8 h of labor per 150 kg of crop residues were estimated for cutting, collecting, loading, and transporting to or from a nearby market.

The comparison of the crop residue application rates demonstrated that the *high mulch* application increased grain production, but if crop residues were monetarily valued, the gross margin per unit labor was lower than for the other uses (Table 4.2–1). The *zero mulch* treatment had the highest remuneration for weeding time. In the case where crop residues were not monetarily valued, returns to land, labor, and weeding were the lowest for the *zero mulch* and the highest for the *high mulch* use.

The gross margins were useful in indicating the returns to land and labor for the crop residues uses with fixed price relations, but were unsuitable for showing the profitability of these uses under farming conditions. The profitability of different crop residue application rates changed, depending on the market value of crop residues (Table 4.2–1). Therefore, the internal value of crop residues, indicated by its shadow price, had to be determined. If mulching reduces benefits derived from other crop residue uses on the farm, the reduction of earnings from these uses (opportunity costs) needs to be compared with the benefits of mulching. Thus, if crop residues are bought (or sold), the purchasing (or selling) price needs to be compared with alternative capital uses for trade to generate income. Moreover, if additional efforts are to be made to increase crop residue production, the costs of land, labor, and capital linked with this increased production have to be considered.

Mulching at a rate of 2000 kg ha^{-1} increased weed infestation and this caused an extra labor requirement during this peak labor period (Lamers and Bruentrup 1996). Given that weeding is already an important constraint in traditional millet production systems in western Niger (Baidu-Forson 1988; McIntire *et al.* 1989), the employment of external labor will increase the costs for *high mulch* and deviate resources from other farm activities. Thus, it appears that it may not be economically desirable for the farm household to devote additional labor to the application of high amounts of crop residues. On the other hand, fodder weeds from the cultivated millet fields may be traded as a valuable feed in Niger (Baidu-Forson 1994). The labor needed to transport crop residues to or from markets should be taken into account if crop residues are actually sold or purchased. For the internal transfer of crop residues to other fields or to other uses, labor efforts are smaller than for marketing. Nevertheless, the value of crop residues is determined by the benefits generated in the resulting farm organization.

To account for the complex linkages, the crop residue uses had to be evaluated in a farming systems framework. The characteristics of the representative household used in this study were based on surveys in western Nigerien villages (Table 4.2–3). Labor could be hired and opportunity costs for family labor could be calculated. Since the trading of crop residues is not always assured in Niger, two scenarios were analyzed. In *scenario 1*, all crop residues had to be produced and consumed on the farm. In *scenario 2*, the buying and selling of crop residues at market prices was assumed and a maximum of 10% of the generated cash income could be used for purchases. In contrast to the partial analysis based on gross margin calculations (Table 4.2–1), an internal transfer of crop residues, from producing to consuming uses, was integrated into the LP analysis. An associated labor requirement of 7 h t^{-1} for transferring crop residues was assumed. The LP model was designed to use household cash income to hire external labor at a total amount not exceeding 25 000 FCFA during peak labor periods. The opportunity costs of capital were not considered in the analysis. It may well be that an alternative use of the involved capital would give higher returns, and would therefore be a more economic use than revealed in the present analysis. The traditional method of capital accumulation of farmers in West Africa is livestock production, which normally has higher returns to capital than investments in agricultural enterprises.

Table 4.2–3. Characterization of a representative farm household in western Niger

Characteristic	Unit	Endowment
Total farm size	ha	10
fallow area:	%	30
Family size	members	11
working members	%	36
Subsistence requirements	kg millet grain per person	200
self-sufficiency level:	%	90
Total working capacity	days per person	200
weeding time capacity:	days per person	45
Livestock possession	TLU*	3

(*Source*: Baidu-Forson 1988; McIntire *et al.* unpublished)
*TLU = Tropical livestock unit

Results of the Linear Programming Models for Mulch Applications

If farmers were restricted to *zero*, *low*, and *high mulch* uses of residues derived from their own farm and to the use of family labor (*scenario 1*), the application of crop residues was not feasible (data not shown). At least 4000 FCFA was needed to invest in hired labor in order to reach the subsistence level of 200 kg millet grain per person. Village surveys revealed that expenditures between 4000 and 13 000 FCFA per household for hiring labor are common (McIntire *et al.* 1989). External labor comprises up to 60% of the total cash expenditures for agricultural related activities in villages of western Niger (Baidu-Forson 1988). The farm organization in the present study employed *low mulch* (two thirds of the cultivated area) and *high mulch* (one third), with an internal transfer of crop residues of over 1 t from the *low* to the *high mulch* use at cash expenditures of 4000 FCFA and greater. Recently, it was reported that farmers in the vicinity of Niamey cut and carried about 20% of their crop residues to markets, through which they gained an additional income (Baidu-Forson 1994). If the grain yield in the present analysis differed by only 23 kg ha^{-1} between the *zero mulch* and *high mulch* use, the *low mulch* use was less profitable than the *zero mulch* use. Consequently, *low mulch* would be replaced by the *zero mulch* application. In this case, *zero mulch* and *high mulch* each occupied 50% of the total crop area. The results showed that the farm organization was very sensitive to changes in grain production caused by a mulch application.

The surface could be expanded until the total farm area of 7 ha was cultivated if the investment in external labor was increased to 23 000 FCFA, which corresponded to 33 man days. In this situation, farm income almost

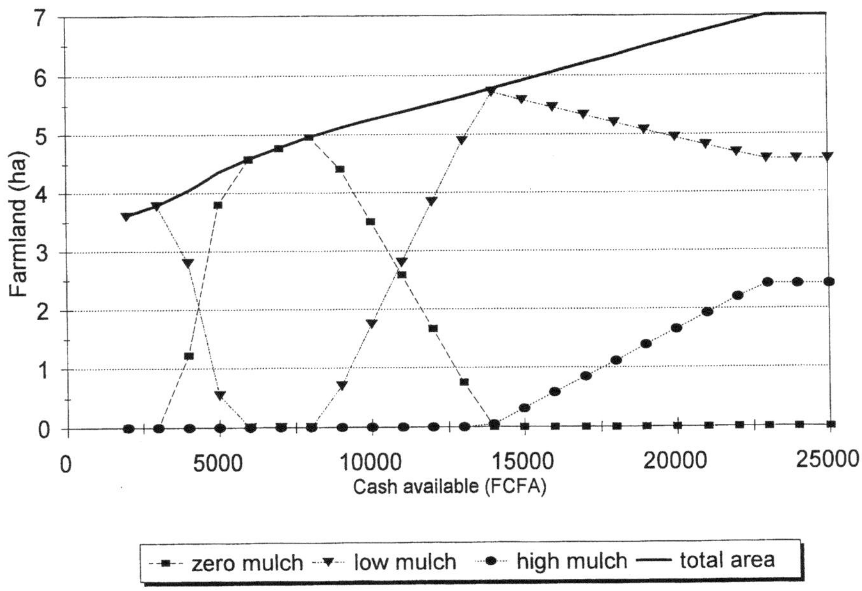

Fig. 4.2–1. Crop residue uses as influenced by cash availability for hiring labor and trading crop residues.

doubled from 176 000 to 326 000 FCFA and cash earnings increased to 150 000 FCFA. The shadow prices for crop residues, the value by which the farm gross margin would change with a one-unit change in the availability of crop residues, varied between 4 FCFA kg^{-1} when labor was scarce and 10 FCFA kg^{-1} when the cultivated area was the limiting factor. The shadow prices were thus less than the market prices (Table 4.2–2).

In *scenario 2*, the optimal combinations of the crop residue uses were more complex (Fig. 4.2–1). To attain the minimum subsistence level, only the *low mulch* treatment was realized because it produced the most millet grain per hour weeding time. However, this use required about 2000 FCFA to hire labor. With increasing cash inputs to hire labor, the subsistence objective could be fulfilled with the *zero mulch* use, thus, replacing the *low mulch* because the *zero mulch* use resulted in higher grain yields per unit weeding time. Moreover, crop residues and, to some extent, millet grain could be sold. If total labor also became restrictive, the *low mulch* would be more profitable than the other crop residue uses due to the high remuneration of total labor. Consequently, the sale of crop residues was reduced and the sale of grain became more important. The *zero mulch* was completely excluded from the farm organization at 14 000 FCFA and replaced by the *high mulch* only. This coincided with an increase in the shadow price of total labor and a decrease in the shadow price for weeding, as well as a reduction of the internal value of crop residues (the shadow price) below 8 FCFA kg^{-1}. Thus, the sale of crop residues decreased in favor of their use as a mulch in high amounts on the field. If land became a limiting factor, only crop residue uses with higher returns to land than the *high mulch* were able to increase the total net revenue. Consequently, the shadow price of crop residues increased compared to the market value of 11.8 FCFA kg^{-1}.

The greatest difference in total farm income between *scenario 1* and *scenario 2* was 23 000 FCFA (a third of the total cash income of *scenario 1*) at a cash expenditure of 8000 FCFA. This difference gradually decreased to zero at cash expenditures of 23 000 FCFA, implying that farmers with low cash availability are tempted to sell crop residues, but at the expense of soil protection and future soil fertility. For *scenario 1*, the ratio of labor investment to total cash earnings decreased from over 30 to 14% with an increased labor availability. In *scenario 2*, this ratio remained between 15 and 12% because crop residues could be sold, thus, increasing cash income more rapidly in situations with low-cash expenditures.

Comparative Advantages of Traditional and Innovative Crop Residue Uses

The results of the LP analysis clearly showed that the optimal use of crop residues depended on a given combination of land and labor constraints, as well as on the monetary resources to relieve them. Labor availability during weeding appears to be especially crucial, confirming earlier results of Baidu-Forson (1988). However, Nigerien farmers use millet crop residues as a multi-purpose commodity (Chapter 4.3). A targeted use of crop residues for controlling wind erosion (Plate 22) would, therefore, compete with these other uses. One cannot expect crop residues to be used as a mulch unless this use is more profitable than the traditional crop residue uses. Therefore, two other uses of crop residues common in traditional farming systems of western Niger were integrated into the model: feeding crop residues to livestock (*crop-livestock*) (Plate 24) and burning crop residues (*burning*) (Plate 26) immediately before the onset of the rainy season.

It was assumed that crop residues were fed to free-grazing animals at a maximum of three tropical livestock units (TLU), consisting of sheep, goats, or cattle, and only during the dry season. To assess the financial value of crop residues as livestock feed, it was furthermore assumed that feeding the livestock with crop residues would stop the live weight losses of Zebu cattle during the dry season. According to Becker and Clar (1991), Zebu cattle could make a live-weight gain of 0.06 kg with 1 kg of crop residues, having a metabolizable energy content of 5.2 MJ kg^{-1} DM. Daily intake for livestock was expected to be satisfied only by millet crop residues and at a maximum rate of 5 kg d^{-1} (Schlecht *et al.* 1995). In the LP model, a feed/conversion ratio (FCR = kg feed per kg live-weight gain) of 25 was used. When livestock is fed with crop residues in the field, manure is produced and distributed by the free-grazing animals. The manure partly replaces the effects of crop residues on soil fertility. The fecal output was calculated based on crop residue digestion as reported by Menke and Steingass (1988). Millet grain and crop residue yields for the *crop-livestock* use were derived from an experiment at the ICRISAT Sahelian Center with animal manure (Ikpe *et al.* 1995). Since manure applications in the experiment were high (1500 kg ha^{-1}), yield adjustments for lower manure rates were estimated for an equilibrium between crop residue production, crop residue consumption, and manure production. The assumed linear increase in millet yields between 0 and 1500 kg ha^{-1} of manure were 40% (grain) and 43% (straw). Total labor time for the *crop-livestock* use was reduced by 7 h ha^{-1} compared to the *zero mulch* use, because crop residues were grazed by animals and subsequently were not cut. No additional labor was considered for herding. Labor time for the first and second weeding operations was expected to be between that of the *zero mulch* and the *high mulch* use, depending on the respective yield levels.

For the *burning* activity, data were taken from a trial at the ICRISAT Sahelian Center where the ash of 2000 kg ha^{-1} of crop residues was applied prior to seeding millet in 1991 and 1992 (Buerkert 1995). Due to the different yield levels between the data sources (for example, the *low mulch* grain yield reported by Buerkert (1995) exceed the *high mulch* yield as presented in Chapter 3.4), the grain and straw yields with *burning* were compared with *low* and *high mulch* yields for the same trial. However, a conversion factor was used to adapt these yields for the gross margin calculations. Since the crop residue balance with *burning* was negative, it was assumed that the additional crop residue demand was purchased to complete 2000 kg ha^{-1}. Labor time for *burning* was the same as the *zero mulch* use, without transport time for selling the crop residues. However, an additional 8 h ha^{-1} was estimated for labor needed to collect and burn crop residues.

The gross margins of the *crop-livestock* and the *burning* uses are summarized in Table 4.2–4. When crop residues were not monetarily valued, both traditional uses showed lower returns to land than the *high mulch* use (Fig. 4.2–2), but higher returns to total labor than all mulch uses (Fig. 4.2–3). *Burning* resulted in the highest grain production per unit weeding time and, consequently, had the highest remuneration per unit weeding (Fig. 4.2–4). When crop residues were valued at market prices (11.8 FCFA kg^{-1}), *burning* had lower gross margins per total labor and per weeding time than the *zero mulch* and *low mulch* uses (Figs. 4.2–3 and 4.2–4). The monetary valuation of crop residues did not change the gross margins for the *crop-livestock* use (Figs. 4.2–2, 4.2–3, and 4.2–4) because the crop residues were directly converted into livestock weight gains. The low labor demand of the *crop-livestock* use is notably important because no material must be transported.

Table 4.2–4. Gross margin[a] calculation for traditional crop residue uses

	Unit	*Burning*	*Crop-livestock*
Input			
Weeding (1st + 2nd)	h ha^{-1}	52	56
Total labor	h ha^{-1}	144	133
Crop residues	kg ha^{-1}	2000	0
Purchase crop residues	kg ha^{-1}	845	0
Manure	kg ha^{-1}	0	558
Total input costs	FCFA ha^{-1}	9970	0
Output			
Total dry matter	kg ha^{-1}	2293	1921
Millet grain	kg ha^{-1}	643	570
Crop residues	kg ha^{-1}	1650	1351
Manure	kg ha^{-1}	0	558
Value of output			
Millet grain	FCFA ha^{-1}	51 432	45 586
Crop residues	FCFA ha^{-1}	0	0
Livestock	FCFA ha^{-1}	0	8476
Total	FCFA ha^{-1}	51 432	54 062
Gross margins with monetary valuation of crop residues			
Per unit land	FCFA ha^{-1}	41 462	54 062
Per unit total labor	FCFA h^{-1}	288	407
Per unit weeding	FCFA h^{-1}	797	970
Gross margins without monetary valuation of crop residues			
Per unit land	FCFA ha^{-1}	51 432	54 062
Per unit total labor	FCFA h^{-1}	357	407
Per unit weeding	FCFA h^{-1}	989	970

[a]Prices for grain, stover, and labor are presented in Table 4.2–2

The gross margin analysis revealed that determining the most beneficial use of crop residues, in terms of gross margins, depended on the labor resources considered and the monetary value of crop residues, making a simultaneous comparison by modeling appropriate. When all alternative crop residue uses on the farm were simultaneously optimized by the LP model, the *crop-livestock* use dominated *scenario 1* at low cash expenditures (Fig. 4.2–5). With increasing cash expenses, the cropping area with *low mulch* and *burning* increased and the surplus of crop residues from the *low mulch* application provided the material for *burning*. *High mulch* entered the optimal farm plan

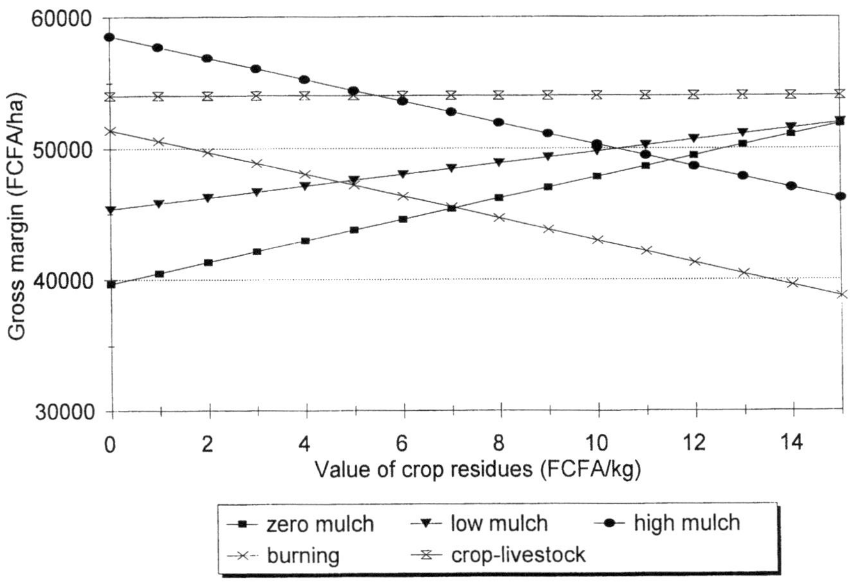

Fig. 4.2–2. Gross margins per ha depending on value of crop residues.

and replaced *burning* when the arable land was limited, which occurred at 20 000 FCFA ha^{-1}. *Crop-livestock* would have dominated all solutions at all levels of cash expenditures if the number of animals possessed by the farm/household had not been limiting (Chapter 1.3). On average, the total income with the *burning* and *crop-livestock* uses was 10 to 15% higher than with mulching techniques only. The shadow price of crop residues (6 to 10 FCFA kg^{-1}) in *scenario 1* was always lower than the market price.

Nevertheless, the optimal solutions in *scenario 1* need to be discussed. For instance, a sensitivity analysis showed that *zero mulch* would replace the *low mulch* use if the gross margin of the former would increase by 500 to 1500 FCFA ha^{-1}, which corresponds to 5 to 12 kg ha^{-1} additional millet grain yield. Also, if the gross margin of *high mulch* would increase by 3000 FCFA ha^{-1} (or 38 kg ha^{-1}), it would enter the optimal solution at cash expenditures lower than 19 000 FCFA. On the other hand, *burning* would always be discarded from the optimal farm organization when land became a constraining resource.

When crop residues were traded (*scenario 2)* (Fig. 4.2–6) and less than 3000 FCFA were available, the use of crop residues was similar to *scenario 1* where trading was excluded (Fig. 4.2–5). However, with a slight increase in cash expenditures for hiring labor, *zero mulch* replaced all other crop residue uses except the *crop-livestock* use. The sale of crop residues generated an additional cash income of about 10 000 FCFA compared to *scenario 1.*

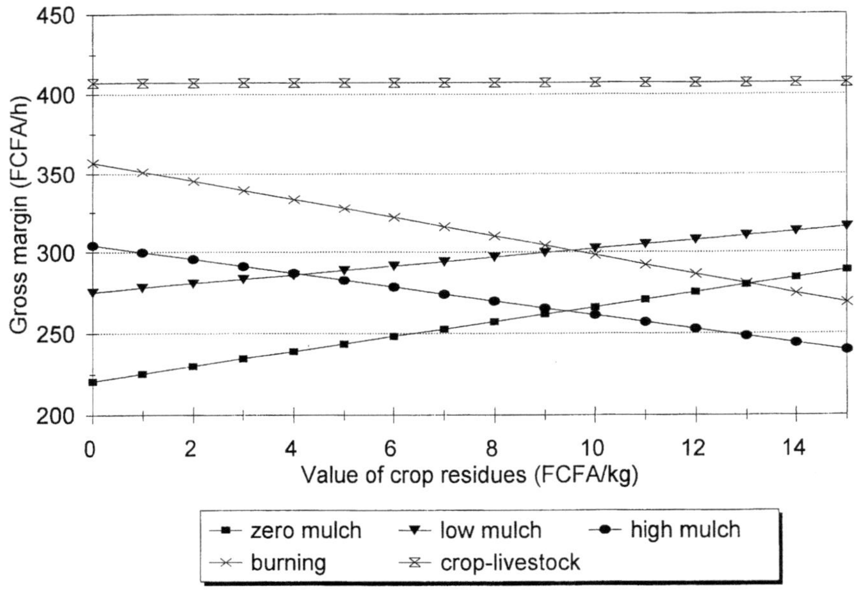

Fig. 4.2–3. Gross margins per total labor time depending on value of crop residues.

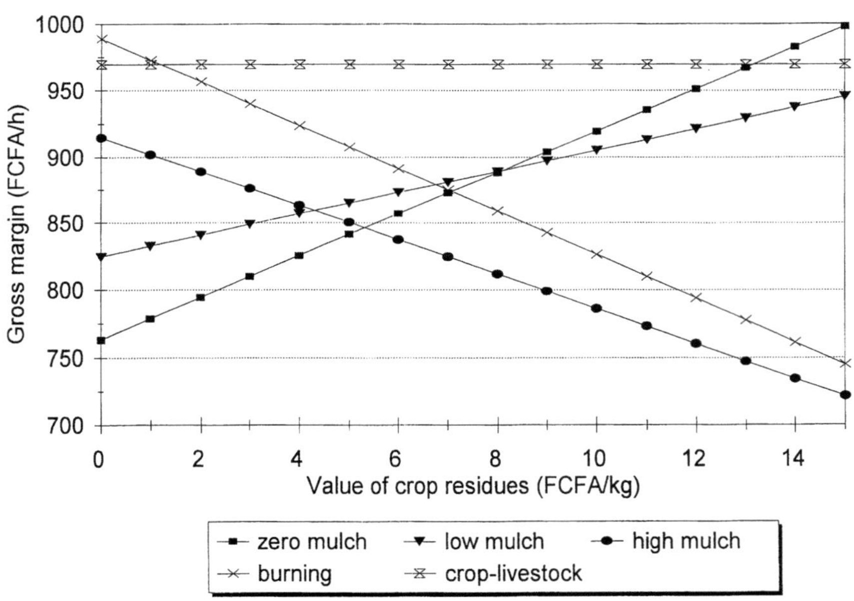

Fig. 4.2–4. Gross margins per weeding time depending on value of crop residues.

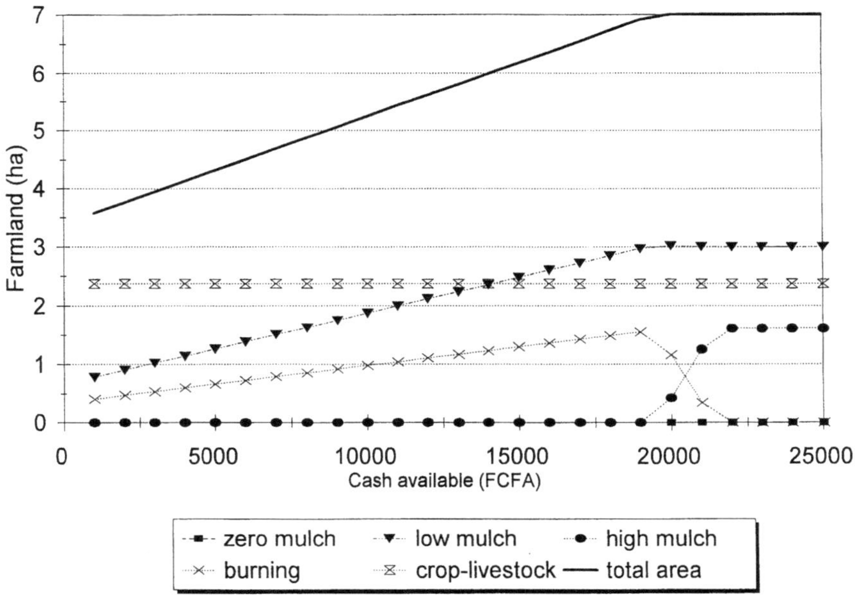

Fig. 4.2–5. Crops residue uses as influenced by cash availability for hiring labor without trading crop residues.

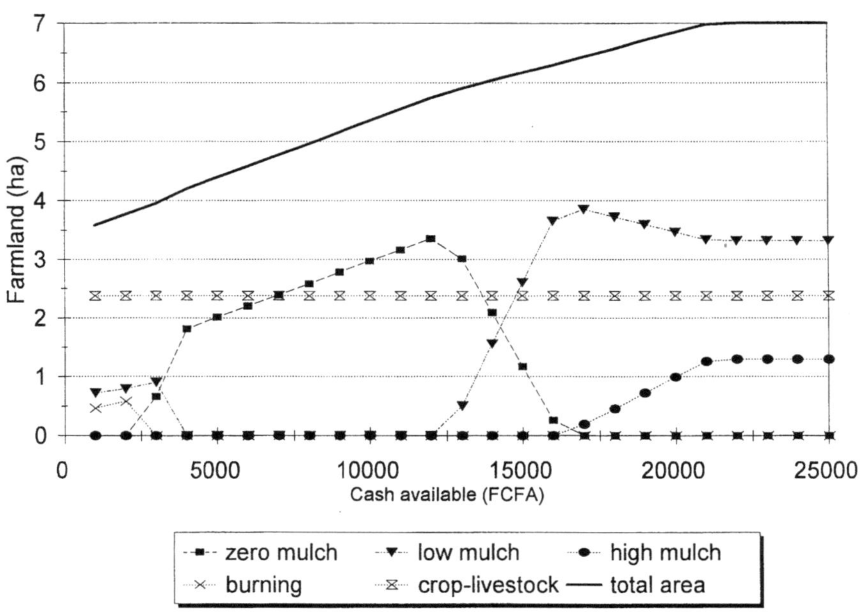

Fig. 4.2–6. Crop residue uses as influenced by cash availability for hiring labor and trading crop residues.

At cash expenditures greater than 12 000 FCFA, the total labor constraint favored the *low mulch* use and, consequently, the sale of crop residues was reduced. Finally, when land became a scarce resource, crop residues were increasingly transferred inside the farm to *high mulch* uses and the solutions under *scenario 1* (Fig. 4.2–5) and *scenario 2* (Fig. 4.2–6) were similar. When crop residues were used on the farm they had an internal value of 6 to 10 FCFA kg^{-1}. The grain production of millet with both mulching techniques needed to be increased by 2000 FCFA ha^{-1} (or 25 kg grain ha^{-1}) for *low mulch* and 5000 FCFA ha^{-1} (or 62.5 kg grain ha^{-1}) for *high mulch* to enter the solution at a point before 12 000 FCFA were spent to hire labor. In *scenario 2*, *burning* was not included in the farm organization if more than 2000 FCFA were spent to hire labor.

The evaluation of the comparative advantage of mulching and the traditional crop residue uses, generally demonstrated that farms with low cash income and severe labor constraints exploited those crop residue uses with the highest returns to labor during weeding to achieve subsistence. In this case, the optimal solutions were mainly a combination of *crop-livestock*, *low mulch*, and *burning*. However, as soon as the self-sufficiency objective was fulfilled, the total labor capacity instead of labor for weeding determined the crop residue uses. Consequently, the most profitable use was now *burning* (to decrease labor demands) or, if possible, selling crop residues. This suggests that when total labor or labor for weeding is a decisive factor for the farm organization, the internal use of crop residues cannot compete with market prices. In contrast, the labor-intensive but high-yielding mulch technique became competitive at 19 000 FCFA or greater for hiring labor in *scenario 1*, and from 16 000 FCFA or greater in *scenario 2*. Even though local farmers are quite knowledgeable about the benefits of using crop residues to enhance crop production, they are reluctant to apply the mulch (Chapter 4.3). This indicates that a typical farm enterprise faces a labor scarcity problem when there is no cash to eliminate this constraint. In addition, the internal exploitation of crop residues as a mulch is inefficient compared to the current uses.

Fertilizer Application

The economic analyses revealed that in the short run the use of crop residues as a mulch was less profitable than other uses known to the Nigerien farmers. However, when the crop residues are continuously exported from the fields, the soil will be susceptible to wind erosion and soil fertility declines. Consequently, a reduction in land productivity can be predicted in the long run. Moreover, it is likely that depressions in crop residue yields make crop residues more scarce and, therefore, more valuable. A crop residue use for mulching at rates of 500 and 2000 kg ha^{-1} will probably become less profitable, because mulching material will need to be purchased. Given this scenario, external crop residue inputs seem to be necessary.

It is well-documented that millet production in the southern Sahelian zone is constrained by the poor inherent fertility of the sandy soils, caused especially by a phosphorus (P) deficiency (Bationo and Mokwunye 1991). Therefore, one possibility for improving the production of millet grain and straw would be the application of P fertilizers. Various experiments at the ICRISAT Sahelian Center showed that the application of P resulted in large increments in millet grain compared to the non-fertilized control (e.g. Bationo and Mokwunye 1991). Recently, Baidu-Forson and Bationo (1992) reported that a combination of mineral fertilizers and animal manure was a more economical management strategy for Sahelian farmers than the use of fertilizers or manure alone. Other experiments revealed that crop residues and P fertilizers have additive effects on dry matter production of millet (Bationo and Mokwunye 1991; Buerkert 1995). Despite the potential for increasing millet yield with mineral fertilizers, the average annual application of 0.3 kg ha^{-1} mineral fertilizer in Niger ranks as one of the lowest in the world (FAO 1993). The objective of the following analysis was to evaluate under what conditions the use of P fertilizers is profitable.

The data were based on experiments where 13 kg P ha^{-1} (applied as single superphosphate at 165 kg ha^{-1}) were added to 500 kg ha^{-1} (*fert500*) and 2000 kg ha^{-1} (*fert2000*) of crop residues. The average absolute increases in grain and crop residue yields were used in the analysis. More details are reported by Buerkert (1995). For the present analysis it was assumed that both crop residues and fertilizer applications would increase labor demands. Data for additional labor linked to fertilizer use (application, weeding, harvest) were taken from an on-farm trial with the same P application rate (Baidu-Forson *et al.* 1994). Sale and purchase of crop residues were associated with 7 h per 150 kg and gross margins were calculated (Table 4.2–5).

Considering the land productivity, the gross margins of *fert500* and *fert2000* were higher than other crop residue uses (Tables 4.2–1, 4.2–4, and 4.2–5). In contrast, returns to total labor and weeding were within or below the gross margins of some other crop residue uses. The LP analyses indicated that without the import or export of crop residues (*scenario 1*), the *fert500* use was only profitable in a land-scarce situation, where it substituted for the *low mulch* use (Fig. 4.2–7). *Burning* was also a profitable alternative with cash expenditures of up to 22 000 FCFA. The *burning* use decreased and *high mulch* became profitable at cash expenses of 22 000 FCFA and greater. Under *scenario 2*, *fert500* entered the optimal farm organization at expenses of 12 000 FCFA (Fig. 4.2–8). Compared to the situation without the purchase of P fertilizers (Fig. 4.2–6), the *high mulch* use was more profitable when fertilizers were included in the farm plan and more than 22 000 FCFA were spent for hiring labor. The results revealed that the additional capital was used to intensify the labor demanding uses rather than to fully exploit the total farm area. Indeed, lack of financial resources, poor access to fertilizers, and an

Table 4.2–5. Gross margins* for fertilizer applications in combination with crop residues

	Unit	*fert500*	*fert2000*
Input			
Weeding (1st + 2nd)	h ha^{-1}	73	82
Labor	h ha^{-1}	208	195
Crop residues	kg ha^{-1}	500	2000
Manure/fertilizer	kg ha^{-1}	165	165
Purchase crop residues	kg ha^{-1}	0	352
Total input costs	FCFA ha^{-1}	10 725	14 873
Output			
Total dry matter	kg ha^{-1}	2865	3362
Millet grain	kg ha^{-1}	842	1007
Crop residues	kg ha^{-1}	2023	2355
Value of output			
Millet grain	FCFA ha^{-1}	67 360	80 560
Crop residues	FCFA ha^{-1}	10 810	0
Total	FCFA ha^{-1}	78 170	80 560
Gross margins with monetary valuation of crop residues			
Per unit land	FCFA ha^{-1}	68 270	66 512
Per unit total labor	FCFA h^{-1}	328	342
Per unit weeding	FCFA h^{-1}	930	807
Gross margins without monetary valuation of crop residues			
Per unit land	FCFA ha^{-1}	57 460	70 660
Per unit total labor	FCFA h^{-1}	276	363
Per unit weeding	FCFA h^{-1}	783	858

*Prices for grain, stover, and labor are presented in Table 4.2–2

insufficient household labor capacity were relevant differences between users and non-users of fertilizers (Thompson and Banaante 1988).

Fert2000 did not enter any solution. A productivity gain of 5000 to 10 000 FCFA ha^{-1} (62.5 to 125 kg ha^{-1} of grain) would be needed for *fert2000* to become economically feasible, as long as weeding is the main constraint and weeding has a high shadow price. When the shadow price of labor decreased (which occurred at cash expenditures of 16 000 FCFA and more), *fert2000* was nearly a competitor. Yet, the *high mulch* use still had an advantage of about 1000 FCFA ha^{-1} (or 12.5 kg ha^{-1} of grain) over *fert2000*. As land became scarce, more crop residues were transferred internally.

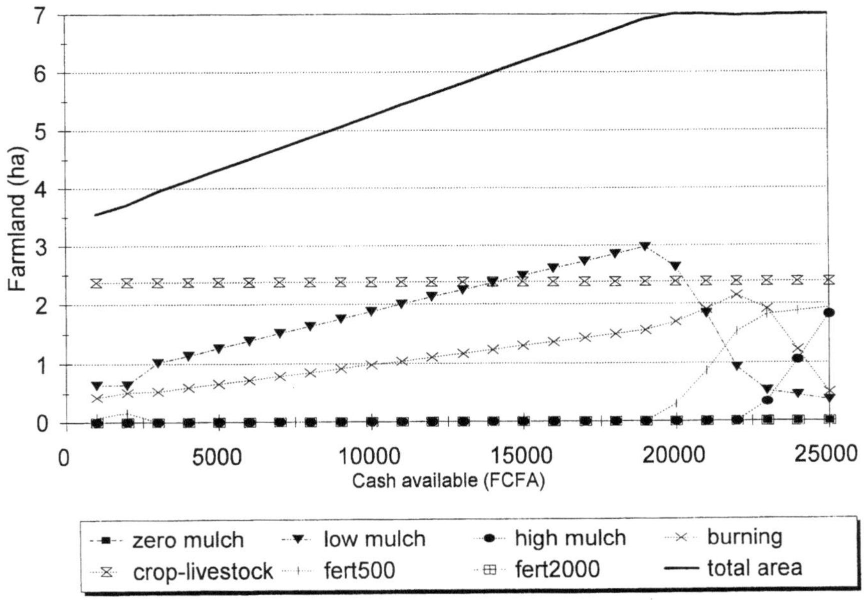

Fig. 4.2–7. Crop residue uses as influenced by cash availability for hiring labor without trading crop residues and with buying single superphosphate fertilizer at 65 FCFA kg^{-1}.

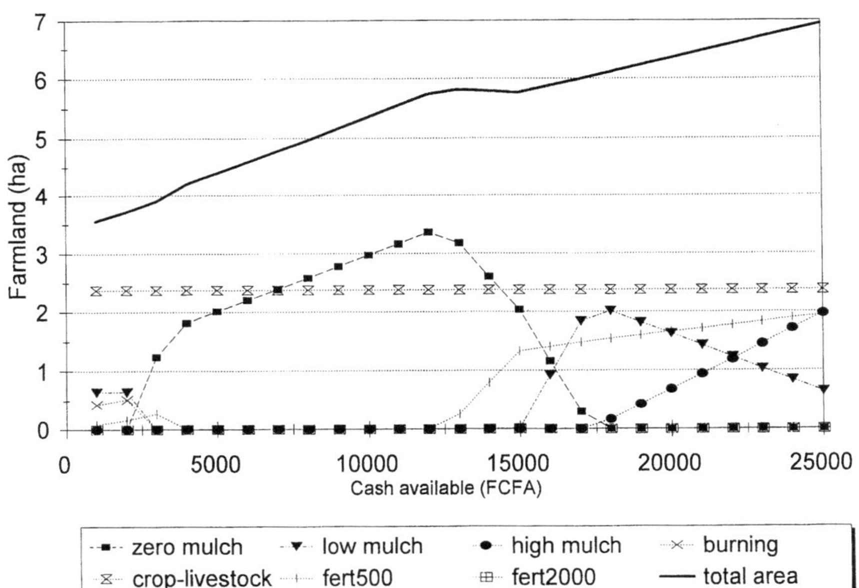

Fig. 4.2–8. Crop residue uses as influenced by cash availability for hiring labor, trading crop residues, and buying single superphosphate fertilizer at 65 FCFA kg^{-1}.

Although agronomic experiments demonstrated the importance of P fertilizers for crop production in the Sahel, the results of the economic analysis clearly showed that the increase in total farm income was minimal compared to the non-fertilizer situations. Using a LP model, Adesina *et al.* (1988) evaluated five new technologies and concluded that farmers' risk aversion, liquidity, and seasonal labor constraints determined their decision-making regarding fertilizer use. At present, the risk involved in investing in fertilizers is not well remunerated by staple crops such as millet. The benefit/cost ratio (BCR), calculated as the additional farm income gained through the use of fertilizers only (benefit) and the additional associated costs of the use of the fertilizer (cost), was less than 1 (results not shown). It is often stated that in developing countries, an investment in fertilizers cannot be expected from farmers if the marginal return barely exceeds the marginal cost. The Food and Agricultural Organization of the United Nations, for instance, estimates a BCR of at least 2 (FAO 1980) for the adoption of fertilizers, indicating that the additional benefits should cover at least twice the additional costs. In the present study, this was true if the fertilizer price was reduced by 50%, assuming that prices for staple crops did not change. At 30 FCFA kg^{-1} single superphosphate and more than 22 000 FCFA available for additional labor, the BCR became greater than 2. Moreover, the investment in land-intensive uses of crop residues, driven by the optimization of net revenues, did not use all the arable surface available to the farmer. The total farm area was cultivated only when 39 000 FCFA were spent for hiring labor (Fig. 4.2–9).

Another strategy for increasing the profitability of P fertilizer applications may be a placed application of low amounts of P fertilizer, for instance, into the sowing pocket. First results showed that even though a placed P application at 0.5 kg ha^{-1} improved early millet growth significantly, dry matter production of millet at maturity was not significantly different from the non-fertilized control (Buerkert 1995). Further research needs to be conducted to evaluate fertilizer placement techniques.

Summary and Final Remarks

The results showed that in western Niger, mulching techniques for wind erosion control were economically feasible under specific conditions only. Given the high expenditures of Nigerien farmers for weeding, they obviously face more labor than land constraints. When the labor constraint is released and land becomes a limiting factor for production, the more labor-consuming mulching techniques (*low* and *high mulch*) become competitive in relation to other more labor-efficient uses of crop residues (*burning*, *zero mulch*, and *crop-livestock*). Farmers without access to cash often sell part of their produced crop residues to achieve household subsistence, particularly when the market price of crop residues exceeds its internal value. This implies that

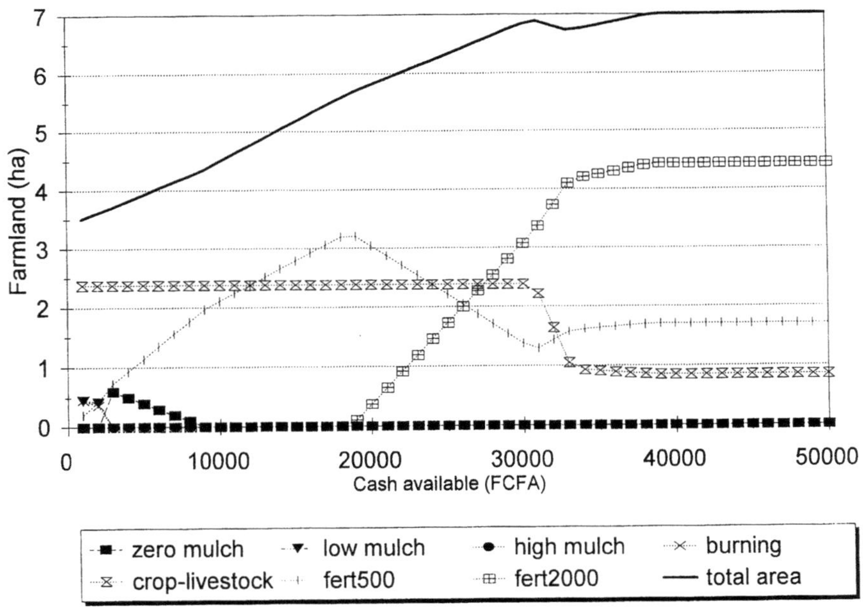

Fig. 4.2–9. Crop residue uses as influenced by cash availability for hiring labor, trading crop residues, and buying single superphosphate fertilizer at 30 FCFA kg^{-1}.

as long as labor is restricting the development of the farm enterprise, these farms cannot fully exploit the benefits of a mulch application. This gives scope for a plea to particularly support resource poor farmers with credit, for hiring manpower, for applying the innovations, and for covering the increased labor requirements associated with mulching and the use of fertilizers.

The economic analysis could not consider the long-term effects of a broadcast mulch application or any other use, as the essential data were not available. A recent study over three years indicated that the practice of burning crop residues does not seem sustainable, for in the third year of burning yields of crop residues and grain declined (Buerkert 1995). This occurred in the absence as well as in the presence of mineral P applied at a rate of 13 kg ha^{-1}. Furthermore, Buerkert (1995) showed that yield depressions in the third year were much smaller with an application of 500 or 2000 kg ha^{-1} of crop residues. Without the use of the mineral fertilizers, the complete removal of millet crop residues for six to eight successive years finally reduced millet grain yields to 39 kg ha^{-1} (Rebafka 1993). The optimal trade-off between short-term income maximization and medium-term fertility management needs further research.

Given the results of the analysis, in the past it was economically more rational for farmers to invest in hiring additional labor than in purchasing fertilizers. Since population growth is expected to continue, a higher use of fertilizers appears to be desirable in the future because less land would be needed for cultivation to fulfill the subsistence objective. Consequently, the over-cultivation of land and soil degradation processes may be slowed down. However, this study clearly showed that the price for P fertilizers determines their economic feasibility and, thus, may influence the adoption of fertilizers. An important research question is whether this price can be reduced, for instance by lowering transport or distribution costs, by the use of local P sources, or by improved technologies for the use of placed P. If a reduction of the fertilizer price is feasible, this would be an important step to stabilize agricultural production. Furthermore, it could be an incentive for farmers to invest in techniques which increase land-productivity, ultimately leading to sustainable systems.

Abstract

Pearl millet crop residues applied in sufficient amounts as a surface mulch reduced soil flux and increased crop yields under the environmental conditions in Niger (Chapters 3.3 and 3.4). However, information about the economics of crop residue uses in Nigerien farming systems is rare. In this chapter, a broadcast mulch of crop residues at rates of 500 kg ha^{-1} (*low mulch*) and 2000 kg ha^{-1} (*high mulch*) was compared with the total removal of crop residues (*zero mulch*) and current practices of farmers burning crop residues (*burning*) or feeding the crop residues to livestock (*crop-livestock*). Due to a lack of long-term data, the analysis only considered the short-term effects of these alternative uses. Gross margins were calculated to compare the profitability of the alternative uses of crop residues which were expressed per unit land area, per unit total labor, and per unit of labor for the two weeding operations. The profitability of these alternative uses was simultaneously evaluated for a representative farm in western Niger with linear programming. The results showed that mulching techniques considered for wind erosion control in western Niger were economically feasible only under specific conditions. Given the high expenditures of Nigerien farmers for weeding, farmers in our study were obviously more constrained by labor than by land. When the labor constraint was released, land became a limiting factor for production. The more labor-consuming but land-saving techniques of mulching became competitive with the more labor-efficient uses of crop residues (*burning*, *zero mulch*, and *crop-livestock*). Farmers without cash could sell part of their crop residues to fulfill the minimum millet consumption demands for the household, particularly when the market price of crop residues exceeded its internal value. The results suggested that as long as labor is restricting the development of a farm enterprise, the full benefits of a mulch application of crop residues cannot be exploited by these farmers. This may be overcome by supporting resource poor farmers with credit for hiring man-power to cover the increased labor requirements associated with mulching and the use of fertilizers. With respect to mineral fertilizers, the analyses indicated that the price for phosphorus fertilizers determined their economic feasibility and, thus, may influence their adoption. A higher use of fertilizers implies that less land needs to be cultivated and thus decreases over-cultivation and, consequently, soil degradation.

Résumé

Une évaluation économique de l'utilisation de résidus de récoltes de mil dans l'optique d'un système de production

Les résidus de récoltes de petit mil appliqués en mulch de surface en quantités suffisantes, réduisent les flux du sol et augmentent les rendements des cultures, dans les conditions d'un environnement naturel au Niger (chapitres 3.3 et 3.4). On a peu d'informations économiques cependant sur les utilisations de ces résidus dans les systèmes de production au Niger. Dans ce chapitre, deux expériences de mulching de résidus de récoltes, l'une à raison de 500 kg ha^{-1} (*bas mulch*) l'autre de 2000 kg ha^{-1} (*haut mulch*), sont comparées avec une situation de suppression totale de mulch (*zéro mulch*) et des pratiques courantes des fermiers qui brûlent les résidus de récoltes (*résidus brûlés*) ou qui nourrissent le bétail de ces mêmes résidus (*culture-bétail*). En raison du manque de données à long terme, l'analyse n'a pris en compte que les effets à court terme de ces usages alternés. Les marges brutes ont été calculées afin de comparer les usages alternés des résidus de récoltes, exprimés par unité de terrain, par unité de force de travail et par unité de travail pour les deux opérations de sarclage. La rentabilité de ces usages alternés était simultanément évaluée sur une exploitation type de l'ouest du Niger par la programmation linéaire. Les résultats montrent que les techniques de mulching sont économiquement réalisables sous certaines conditions seulement. Etant donné les dépenses importantes que les paysans nigériens allouent au sarclage, les agriculteurs de notre étude étaient plus limités par le travail que par la terre. Lorsque la contrainte du travail a été supprimée, la terre est devenue un facteur limitant pour la production. Les techniques de mulching exigeant plus de travail mais économisant la terre se sont avérées concurrentielles avec les usages des résidus de récoltes plus efficaces concernant le travail (*brûlis*, *zéro mulch* et *culture-bétail*). Il arrive même aux agriculteurs sans ressources de vendre une partie de leurs résidus de récoltes pour parvenir à couvrir les besoins minimaux de consommation de petit mil du ménage, en particulier lorsque le prix du marché des résidus de récoltes est supérieur à sa valeur interne. Ceci implique que tant que la disponibilité de travail limite le développement des entreprises rurales, celles-ci ne peuvent tirer profit de tous les bénéfices de l'application de résidus de récoltes. On peut surmonter cette situation par un soutien aux agriculteurs pauvres en leur accordant un crédit pour acheter la force de travail nécessaire aux techniques de mulching et à l'application d'engrais. En ce qui concerne les engrais minéraux, les analyses indiquent que le coût de la fertilisation en phosphore détermine leur profitabilité économique et est donc décisif quant à leur adoption. Un plus grand usage des engrais entraîne un besoin moindre en terre cultivée évitant ainsi sa surexploitation et donc aussi la dégradation du sol.

References

Adesina, A.A., P.C. Abbott, and J.H. Sanders. 1988. Ex-ante risk programming appraisal of new agricultural technology: Experiment station fertilizer recommendations in southern Niger. *Agricultural Systems* 27: 23–34.

Baidu-Forson, J. 1988. Characteristics of farm households and economy in western Niger villages: Some evidence from baseline surveys. Resource Management Program Economics Group Progress Report 87, International Crops Research Institute for the Semi-Arid Tropics, Niamey, Niger.

Baidu-Forson, J. 1994. Uses and economic value of crop residues and herbaceous plants to rural households in western Niger. Resource Management Program, Progress Report No. 5, ICRISAT Sahelian Center, Niamey, Niger.

Bationo, A. and A.U. Mokwunye. 1991. Role of manure and crop residues in alleviating soil fertility constraints to crop production: With special reference to the Sahelian and Sudanian zones of West Africa. *Fertilizer Research* 29: 117–125.

Baidu-Forson, J. and A. Bationo. 1992. An economic evaluation of a long-term experiment on phosphorus and manure amendments to sandy Sahelian soils: Using a stochastic dominance model. *Fertilizer Review* 33: 193–202.

Baidu-Forson, J., C. Renard, N. Deuro, C. Reddy, and M. Kadi. 1994. Evaluation agro-économique de nouveaux systèmes de production en milieu paysan au Niger. Programme d'Exploitation des Ressources, Compte rendu des travaux No. 4, ICRISAT Sahelian Center, Niamey, Niger.

Becker, K. and U. Clar. 1991. Potential of animal draught power – Limitations and prospects. *Animal Research and Development* 34: 9–16.

Buerkert, A. 1995. Effects of crop residues, phosphorus, and spatial variability on yield and nutrient uptake of pearl millet (*Pennisetum glaucum* L.) in southwest Niger. Diss. Universität Hohenheim, Verlag Ulrich E. Grauer, Stuttgart, Germany.

FAO. 1980. Les engrais et leurs applications. Précis à l'usage des vulgarisateurs. Collection FAO: Mise en valeur des terres et des eaux, No. 8, 3ème édition, FAO, Rome, Italy.

FAO. 1993. AGROSTAT-PC. Computerized Information Series. Statistics Land Use 1/3, FAO, Rome, Italy.

Ikpe, F.N., J.M. Powell, and N.O. Isirimah. 1995. Primary tillage and nutrient cycling in mixed farming systems of semi-arid West Africa (SAWA). pp. 410–437. *In* G. de Noni, J.F. Nouvelot, and E. Roose (eds.) Réseau Erosion Bulletin 14. Proceedings Workshop on Soil Erosion in Montpellier, 15–18 Sept. 1993, ORSTOM, France.

Lamers, J.P.A. and M. Bruentrup. 1996. Comparative advantage of single and multipurpose uses of millet stover in Niger. *Agricultural Systems* 50(3): 273–285.

McIntire, J., J. Hopkins, J.P.A. Lamers, and L.K. Fussell. 1989. The millet system of western Niger. I. Crop production. ILCA, Addis Abeba, Ethiopia (unpublished).

Mellor, J. and B.F. Johnson. 1984. The world food equation: Interrelation among development, employment and food consumption. *Journal of Economic Literature* 22: 532–574.

Menke, K.H. and H. Steingass. 1988. Estimation of the energetic feed value obtained from chemical analysis and in-vitro gas production using rumen fluid. *Animal Research and Development* 37: 7–55.

Ministère des Ressources Animales. 1991. Rapport statistique mensuel sur le suivi des marchés à betail. Projet Collecte d'Information sur les Marchés, Niamey, Niger.

OPVN. 1992. Marchés des céréales et du Niébé au Niger 1990/1992. Office des Produits Vivriers du Niger: Système d'Information sur les Marchés (S.I.M.), Niamey, Niger.

Rebafka, F.P. 1993. Deficiency of phosphorus and molybdenum as major growth limiting factors of pearl millet and groundnut on an acid sandy soil in Niger, West Africa. Hohenheimer Bodenkundliche Hefte, No. 9, Stuttgart, Germany.

Schlecht, E., F. Mahler, M. Sangaré, A. Susenbeth, and K. Becker. 1995. Quantitative and qualitative estimation of nutrient intake and faecal excretion by Zebu cattle grazing natural pasture in semi-arid Mali. pp. 85–97. *In* J.M. Powell, S. Fernández-Rivera, T.O. Williams, and C. Renard (eds.) Livestock and sustainable nutrient cycling in mixed farming systems and sub-Saharan Africa. II. Technical papers. Proceedings International Conference, Addis-Ababa, Ethiopia, 22–26 Nov. 1993, ILCA (International Livestock Centre for Africa), Addis Ababa, Ethiopia.

Thompson, T.P. and C.A. Banaate. 1988. A socio-economic study of farm – level constraints to fertilizer use in western Niger. IFDC Paper Series P-6, IFDC, Muscle Shoals, Alabama.

4.3 Farmers' Perceptions about Wind Erosion and Its Control

Petra Feil and John P.A. Lamers

Introduction

Two household surveys were conducted in western Niger to complement the on-station wind erosion experiments that focused on the use of crop residues (Chapters 3.3 and 3.4) and windbreaks (Chapters 3.1 and 3.2) for wind erosion control. The objective was to learn about the awareness and the state of the farmers' knowledge about wind erosion and indigenous erosion control techniques. Agricultural researchers have examined the use of millet crop residues to reduce wind erosion, therefore the role of this by-product within the farming systems was assessed. A second survey was associated with the planted tree windbreak experiment (Chapter 3.2). The uses and appreciations of local tree and shrub species within the existing farming systems were evaluated.

Survey Methodology

Both surveys were conducted in 1992 in the villages of Djiakindi and Sadoré. Both villages are located near the ICRISAT Sahelian Center, 45 km south-east of Niger's capital, Niamey. The survey was based on an open guideline interview and followed farmers' reasoning to grasp their perceptions and to relate these to their explanations and concepts. Interviews were conducted in Djerma/French with the help of an experienced translator. In the first survey, 29 farmers volunteered to participate and were interviewed. In the second survey, 21 farmers and 20 of their wives agreed to participate, while four wives refused.

B. Buerkert, B. E. Allison and M. von Oppen (eds.), Wind Erosion in Niger, 215–231.

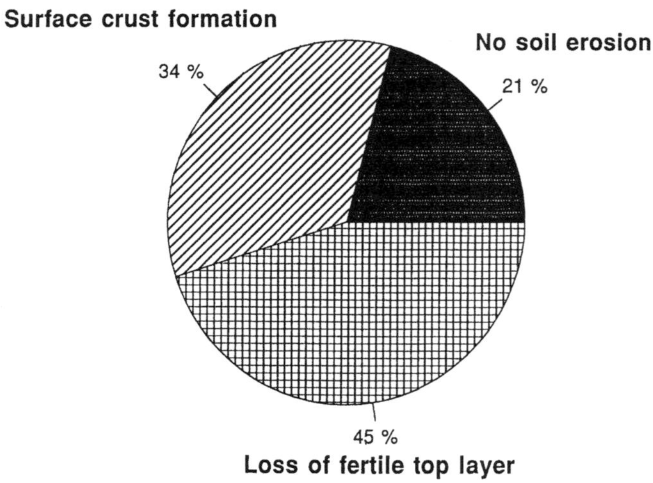

Fig. 4.3–1. Effects of wind erosion on soils, according to farmers in western Niger in 1992 (n = 29).

Local Knowledge of Wind Erosion and Its Relation to Soil Fertility

Wind Erosion

Farmers confirmed that wind-generated soil erosion was on-going. Soil losses and injury to millet plants were considered as the principle damage from wind erosion. Although 66% of the farmers responding declared that wind erosion damage occurred on their fields within the last decade, it was not regarded as a major concern except on specific spots (Fig. 4.3–1). The loss of the surface soil and the subsequent development of a surface crust were mentioned as damage to the soil. Direct sandblast damage to the millet crop was recognized by 69% of the farmers responding (Fig. 4.3–2). High-speed storm events caused no crop damage according to 31% of the farmers. During the last ten years, farmers referred only to the abnormally dry year 1984 (less than 250 mm total rainfall) when severely damaging wind erosion events occurred. Despite several replantings, yields were extremely low in that year (less than 100 kg ha^{-1}). Yet, this yield reduction was attributed to the low precipitation rather than to wind erosion.

At the onset of the growing season, all farmers stated that the coverage of young millet seedlings by wind-blown material was the most severe damage caused by storms. Seven farmers mentioned that such damage occurred at a stage when millet seedlings had less than three leaves. The other 22 farmers did not specifically mention the growth stage of millet. The sensitivity of young millet seedlings to coverage with soil in the early stages of development was confirmed by field and wind tunnel experiments (Chapter 2.2). When

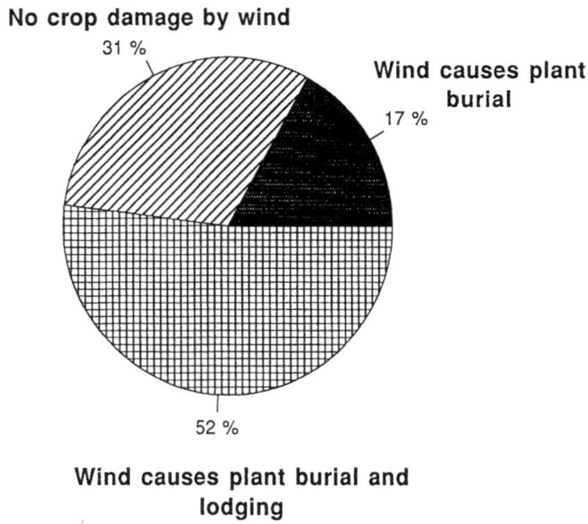

Fig. 4.3–2. Effects of wind erosion on plants, according to farmers in western Niger in 1992 (n = 29).

millet was manually covered with soil during three different growth stages, the survival at the 1-leaf stage was 28% higher than burial at the 2- or 3-leaf stage. Above-ground dry matter losses of 66% were recorded when millet seedlings were completely covered by soil. The delay in millet growth and development after being exposed to wind was also mentioned by farmers. However, they stated that millet plants would die only when the plant burial with wind-blown soil was followed by drought and high temperatures. These factors have been discussed by researchers (Michels 1994).

According to 52% of the farmers, lodging caused by strong winds at the end of the growing season was equally as damaging as burial at the seedling stage (Fig. 4.3–2). Even though they stated that lodging delayed ripening, the actual loss was caused by termites and mice which have easy access to lodged millet heads. However, scientific evidence on lodging damage due to wind is lacking.

Farmers' Knowledge on the Relation between Wind Erosion and Soil Fertility

Farmers' fields in the survey region were characterized by uneven surfaces with small ridges and depressions. Of the 29 farmers interviewed, 48% allocated different soil fertility levels to these micro-topographical features. The micro-highs (*fandus* in Djerma) or ridges were judged to be the fertile parts, whereas micro-lows (*gorous*) or depressions with a surface crust (*gaganis*) were considered to be less fertile. Knowledgeable farmers explained further that barriers such as natural vegetation or dead twigs, branches, and tree stumps capture wind-blown soil particles. Soils with the accumulated sandy top layer retain more water than soils with thinner sandy top layers. The vegetation is considered to be more dense on *fandus* than on *gorous*. Through termite activity or hoeing, the organic material consisting of leaves, branches of trees and shrubs, or crop residues is mixed with the topsoil layer. This augments the organic matter content of the *fandus* and, consequently, the soil fertility. Recently, the relationship between soil fertility and micro-physiography was confirmed by Geiger and Manu (1993). Millet grain yields on sites with micro-highs, comprising 30% of the planted millet surface, were 120% higher than on sites without micro-highs. Even though farmers provided an explanation for the better millet growth on sites with *fandus*, researchers have not yet found reasons for it (Geiger and Manu 1993).

Farmers considered the topography and the decreasing soil fertility as two interacting factors that have led to field spots with increased wind erosion damage. The relationship between decreasing millet yields, due to low soil fertility or inadequate rainfall, and wind erosion events was noted by 48% of the responding farmers. They indicated that a reduced amount of crop residues made the topsoil (particularly in micro-lows) more susceptible to wind erosion. Therefore, soil erosion decreased the soil fertility which, subsequently, caused even greater yield reductions. Also, the loss in vegetative cover leaves the surface more susceptible to erosion. This process was noticed especially during the dry season which is characterized by a low vegetative cover and 'Harmattan' winds (Plate 5). If the erosion process is not hindered, the soil will eventually be replaced by a thin surface crust. This red denuded soil crust is named *gangani* (or *gangani kirey*). The surface crust is characterized by low infiltration rates and a low fertility status. If the erosion process continues and the *gangani* extends in area, the *gangani* is called *tanka*. *Ganganis* and *tankas* were regarded as the predictable final results of the erosion process.

Hoogmoed and Stroosnijder (1984) confirmed that the sandy soils predominating in the West African Sahel are very sensitive to crust formation. A survey classified *gorous* as eroded surfaces caused by sheet erosion, and *ganganis* as exposed B_t horizons after the removal of the A horizon (Taylor-Powell *et al.* 1991). The existence of special terms in Djerma for the different stages of soil surfaces subjected to wind erosion is strong evidence that farmers are quite knowledgeable about the wind erosion process. However, although farmers stated that wind erosion mainly occurs during the 'Harmattan' season, Michels (1994) found that erosion during this entire 3-month period did not exceed the erosion that took place during one storm event in the rainy season.

Farmers' Use of Crop Residues in Wind Erosion Control

Farmers were not only aware of the wind erosion process but they also made efforts to restore eroded spots. Although the surface crusts of *ganganis* can be destroyed by manual hoeing, farmers preferred another less labor intensive solution. Their technique aimed first at retrieving a new soil layer and then at improving the soil fertility status during the rehabilitation process. The *ganganis* (*tankas* are normally left untouched) were mulched with high amounts of millet crop residues, visually estimated at 0.4 to 0.6 kg m^{-2} (Plate 23). The material was taken from the more productive spots or from trees and shrubs. A recent study showed how mulching with branches from trees and shrubs stimulated re-vegetation on bare, over-exploited Sahelian forest soils in Niger (Chase and Boudouresque 1987). The applied mulch captured soil particles and seeds of herbaceous plants and trees. During the first rainy season the vegetation covered the ground immediately and the surface crust slowly disappeared.

The trapping of aeolian soil material usually takes one to two years, according to 48% of the surveyed farmers. It was discussed earlier (Chapters 3.3 and 3.4) that a broadcast surface mulch at a rate of 2000 kg ha^{-1} of crop residues reduced soil flux by about 50%. The total soil flux at 0.1 m height above ground level was more than 1260 kg m^{-2} during one rainy season (Chapter 2.2). Detailed topographic studies confirmed that soil erosion caused by wind and water can be effectively reduced by crop residues. Additionally, the surface level increased with the application of crop residues due to soil deposition (Geiger *et al.* 1992; Chapter 3.3). In another experiment at the ICRISAT Sahelian Center, the perennial grass *Andropogon gayanus* reduced soil erosion and trapped 2000 t ha^{-1} of soil within three years (Renard and Vandenbeldt 1990).

During the restoration process of *ganganis*, farmers envision first of all the improvement of the soil physical parameters and, subsequently, the soil chemical features. Mulch captures the aeolian material and at the same time the mulch is decomposed and hence, increases soil fertility. Within one growing season, about 70% of the total dry matter of residues is gradually decomposed due to termite and microbial activity (Hebel, personal communication). Mulching with 2000 kg ha^{-1} of millet crop residues increased the availability of calcium, magnesium, and the pH at 0 to 0.01 m soil depth (Chapter 3.4).

Despite the obvious yield increases, farmers (41%) acknowledged the importance of crop residues more as a means for trapping aeolian material and less for enriching the soil directly. Powell and Ikpe (1992) reported that millet crop residues initially immobilized soil nitrogen. They concluded that millet crop residues caused nitrogen deficiency during the early stages of millet growth, and that immobilized nitrogen was released at a growing stage when the plants could not benefit anymore.

In many countries in sub-Saharan Africa, indigenous measures were developed to maintain soil fertility. These include mechanical (terraces, contour drains, earth and stone bounds, contour rows of plant debris, tie-ridging), biological (mulching, cover cropping, planted contour lines, agroforestry, incorporation of organic matter, green and animal manure, crop rotation, and leguminous crops), and institutional arrangements (herding strategies, irrigation management) (Hailu and Runge-Metzger 1992). The authors concluded that the existence of indigenous practices was no assurance that they were effective in establishing sustainable agro-ecosystems. Moreover, soil degradation is a very slow process and although farmers observed this process, they mainly attributed the reduction in crop yields to declining rains. As long as farmers do not perceive soil degradation as a major factor limiting their yields, the adoption of sustainable soil conservation techniques on a large scale is not very likely (Shaxson 1985).

Recently, it was argued that the soil micro-variability, a typical feature of the soils in the Sahel, may be an asset to the subsistence-oriented farming systems because it could reduce the risk of poor harvests (Brouwer *et al.* 1993). Other researchers have recommended adapting the management of the micro-lows and micro-highs more to their specific conditions, assuming that the sites reacted differently to inputs such as fertilizers (Geiger and Manu 1993). In the first two years (after a fallow of nine years), a crop residue application of 2000 kg ha^{-1} increased millet straw and grain yields on initially poor producing spots by 42 and 48%, respectively, compared with an application of 500 kg ha^{-1} (Buerkert *et al.* 1995). On initially good producing spots, only grain yields increased (9%). Given these yield differences, the marginal net return from a concentrated crop residue application on poor producing spots is higher than the marginal net return from a broadcast crop residue application. Yet, recently it was argued that a broadcast application of 2000 kg ha^{-1} is not sustainable (Feil *et al.* 1995). Moreover, a broadcast mulch of millet crop residues at 2000 kg ha^{-1} stimulated not only millet growth, but also weed production. Consequently, labor time for weeding operations increased by 25 to 30% (Chapter 4.2). Weeding is already the most labor demanding activity in millet production (Baidu-Forson 1988) and an increase in weeding time might be an additional reason for farmers' preferences. Labor scarcity during the cropping season might explain why farmers want to have a 'clean field' at the onset of the rains. They not only decide against a broadcast mulch with crop residues, but even burn them at the end of the dry season (Plate 26). The widespread practice of burning could also be a carry-over from the traditional slash and burn cropping system. Burning crop residues before the growing season is also the recommended remedy against an infestation of the stem borer (*Coniesta* (*Acigona*) *ignefasalis*), since insecticides are not available. Even though burning millet crop residues at a rate of 2000 kg ha^{-1} showed similar yield increases to a broadcast application with the same quantity, burning does not seem to be a sustainable practice in the long run because yields decreased in the second year after burning (Rebafka 1993).

Millet crop residues are used by Nigerien households for quite diverse purposes (Fig. 4.3–3) and this limits its availability for mulching. Depending on the objective of the farmer, the millet crop residues are harvested or left in the field for soil protection or animal feed (ICRISAT 1992; Powell and Ikpe

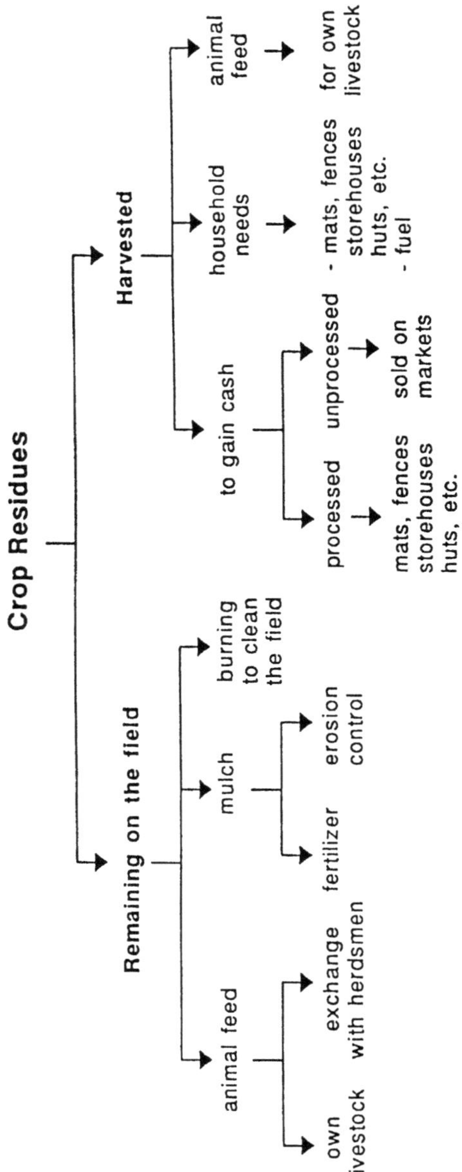

Fig. 4.3–3. The uses of millet crop residues by households in western Niger in 1992.

1992). Livestock owners take their animals to their harvested millet fields to graze during the day (Plate 24). During the night, livestock is gathered to manure on selected fields or selected spots within one field (Powell and Ikpe 1992). Thus, livestock transfer fertility from one place or area to another. According to a recent study in western Niger, most residues disappeared in the first two months after the grain harvest, although millet leaves diminished quicker than stalks (ICRISAT 1992). During the dry season, the low quality millet stalks represent a precious livestock feed because better alternatives are rarely available.

In western Niger, a small percentage of millet stalks are harvested for constructing sheds, fences (Plate 25) for gardens, compounds or showers, grain stores, mats for a hangar or for animal enclosures, for use as fuel, or for sale (Gavian 1992). With the gradual loss of the rangelands, farmers have become aware of the rising market value of crop residues, particularly in the vicinity of urban centers (Speirs and Olsen 1992). In 1992/1993, harvested crop residues were sold at markets in Niamey and Djiakindi for about 7 to 10 FCFA kg^{-1} (Chapter 4.2). These modest earnings can be important during years of poor grain harvest. Recent studies showed that in the proximity of Niamey, Niger, about 20% of the millet crop residues were cut and carried from the field (ICRISAT 1992). In central and south Niger, the quantities of harvested crop residues were higher, due to their use as fuel and animal feed, leaving the soils bare and susceptible to the eroding 'Harmattan' winds (Angstreich 1991).

Local Knowledge of Trees and Shrubs

Appreciation of Trees and Shrubs

The survey showed that men and women viewed the importance, appreciation, and reasons for appreciation of local trees and shrubs differently (Table 4.3–1). Farmers judged tree and shrub species with respect to their role in agriculture, such as the enrichment of soil or the prevention of soil erosion. Their wives appreciated tree species for the provision of firewood (Plate 17), human consumption, or shade for resting. Both farmers and their wives agreed on the importance of the browse characteristics for animal feed. Thus, the importance of the species seems to be related to gender. A recent survey in Kenya revealed that the choice of tree species, the tree product marketing and use, and the employment of soil conservation and fertility management practices were not only differentiated by gender, but also by the marital status of the women (Bonnard and Scherr 1994). The reasons for the 'dislike' of tree and shrub species by farmers were also related to agricultural issues (Table 4.3–2). The reasons most mentioned were the attraction of birds and insects, or the provision of too much shade, hindering millet growth. The surveyed wives

Table 4.3–1. Reasons for appreciating local trees and shrubs (24 species) by farmers (n = 21) and their wives (n = 20) in western Niger in 1992

Reasons for appreciation	Farmers		Wives	
	Rank	Frequency (%)	Rank	Frequency (%)
Soil enrichment	1	45	–	n.m.
Erosion protection	2	20	–	n.m.
Animal feed	3	17	2	25
Shade	4	7	4	17
Construction material	5	3	6	<1
Human consumption	6	1	3	23
Medical	6	1	5	5
Firewood	6	1	1	28
No special reason		5	–	2

(*Source:* Lamers *et al.* 1994a)
n.m. = not mentioned

Table 4.3–2. Reasons for disliking trees and shrubs (24 species) by farmers (n = 21) in western Niger in 1992

Reasons for dislike	Rank	Frequency (%)
Abundant shade reduces millet growth	1	17
Attracts birds	2	15
Attracts insects	3	14
Enhances erosion	4	11
Depletes soil of minerals and soil moisture	5	6
Human nuisance (thorns, etc.)	6	3
Slow growth rate	7	1
No special reason	–	33

were not interested in this question; therefore, the limited responses were not included.

Trees and Their Impact on Wind Erosion

During the discussions on wind erosion, farmers repeatedly reported the negative or positive impact of certain tree and shrub species on the wind erosion process (Fig. 4.3–4). Tree species such as *Combretum glutinosum*,

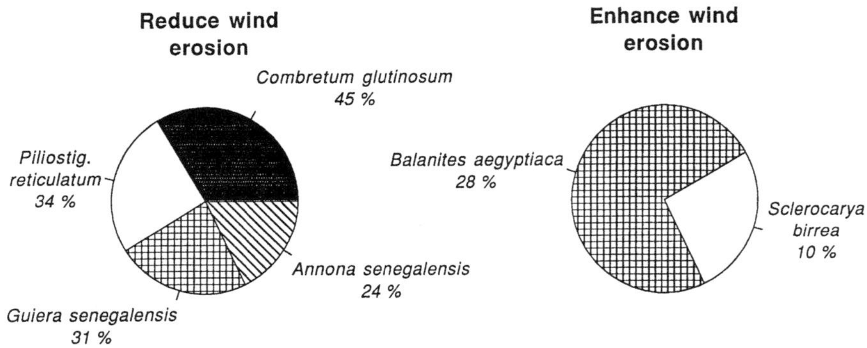

Fig. 4.3–4. Tree and shrub species considered to enhance or reduce wind erosion, according to farmers in western Niger in 1992 (n = 21).

Piliostigma reticulatum, *Guiera senegalensis*, and *Annona senegalensis* were considered by farmers to be effective against wind erosion. Their observations were recently confirmed by Geiger and Manu (1993), who found that the better producing *fandus* in farmers' fields were associated with the growth of tree species in general, and in particular with *G. senegalensis*, *C. glutinosum*, or *P. reticulatum*. Farmers noticed increased wind erosion damage with trees such as *Balanites aegyptiaca* or *Sclerocarya birrea*. As these trees grow taller, wind erosion increases directly around the stem.

It is known that the characteristics of air-flow are dominated by the geometry of the obstacle and that the impact of the wind speed depends on the stem height (Gross 1987). Although individual trees reduce wind speed, and thus wind erosion, their effect is lower than a combination of scattered trees, which is usually found on farmers' fields (Kainkwa and Stigter 1994). Nevertheless, farmers in Niger tolerate species which enhance erosion because they produce fruits and leaves for human consumption, as well as for animals. None of those interviewed had ever planted or cut any tree species to specifically control wind erosion.

In evaluating the windbreak species included in the on-station windbreak experiment (Chapter 3.2), 52% of the farmers assessed *Azadirachta indica* windbreaks not to be effective, or only fairly effective against wind erosion. This contradicts, in part, the data that has been reported on the *A. indica* windbreaks in the Majjia Valley in central Niger (Plate 16). These windbreaks reduced wind speeds up to 50% and, consequently, wind erosion (Long and Persaud 1988). Yet, the millet yield increase of 26% compared to fields without windbreaks, was not statistically significant. Moreover, yield reductions were reported with the same *A. indica* windbreaks in other years (Kerkhof 1991). Despite the high appreciation of *F. albida* by the rural people in Niger, 38% of the farmers considered this species to be only moderately effective against wind erosion. In contrast, *B. rufescens*, which was less appreciated

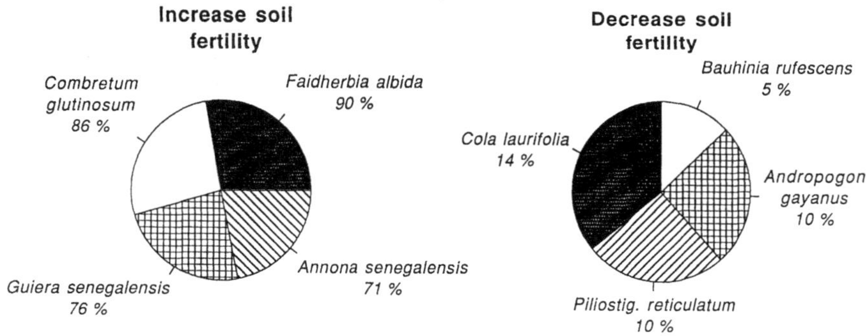

Fig. 4.3–5. Tree and shrub species considered to increase or decrease soil fertility, according to farmers in western Niger in 1992 (n = 21).

by farmers than many other species, was considered by 90% of the farmers to be very effective against wind erosion.

Trees for Soil Fertility and Soil Conservation Management

Some tree species and shrubs were considered to maintain, improve, or conserve soil fertility, while others were thought to deplete the soil (Fig. 4.3–5). Farmers knew that millet grows faster and taller underneath and around *F. albida* compared to the surroundings (Plate 8). They explained this improvement, in part, to the presence of animals which are attracted by the highly palatable pods and leaves of *F. albida*, and subsequently deposit manure while grazing. At a recent workshop on *F. albida*, evidence was summarized on the side effects of this species on microclimatic factors, soil properties and crop yields (Sall 1992). The author concluded that underneath *F. albida*, total soil carbon and nitrogen was increased by 40 to 100% and biological activity two to fivefold. The litter drop combined with high biological activity in the soil was considered as the main soil improving effect of this species (Sall 1992), although the crude protein content of *F. albida* leaves was the second lowest among the six tree windbreak species (Lamers *et al.* 1994b).

Millet grain yields with *F. albida* windbreaks increased by an average of 17% (Chapter 3.2), even when manure was excluded from these experiments. The capacity of this species to fix nitrogen may have caused the increased soil fertility, although *F. albida* had the lowest N_2 fixation among eleven woody and eleven non-woody Mimosaceae (Schulze *et al.* 1991). Another experiment showed that soil temperatures underneath *F. albida* were closer to the optimum for millet growth and development, which allowed the better plant growth (Vandenbeldt and Williams 1992). Even though researchers and farmers have observed yield increases underneath *F. albida*, the explanations are not conclusive.

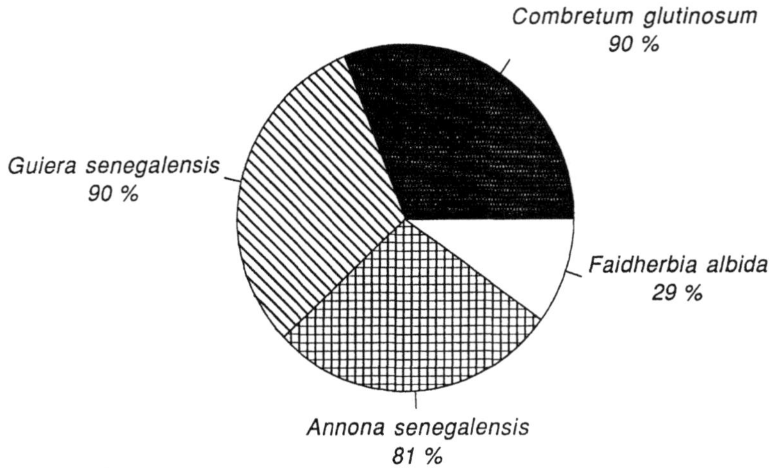

Fig. 4.3–6. Trees and shrubs considered to be good firewood species, according to women in western Niger in 1992 (n = 20).

According to farmers' experience, shade suppresses the growth and development of millet, and *F. albida* was therefore appreciated even more because of its inverted phenology. It is one of the rare tree species in the Sahel that sheds its leaves during the rainy season and is green during the dry season, providing shade for livestock during the hottest part of the day. *C. glutinosum* and *Ziziphus mauritania* were considered to have the same 'fertilizing' effect as *F. albida*, but due to shading, millet can only profit when the trees are heavily trimmed, which is done during the first weeding. *G. senegalensis* and *A. senegalensis* shrubs were appreciated as species for catching aeolian material which subsequently improved soil fertility (Fig. 4.3–5). Farmers stated that the number of several species, such as *A. senegalensis*, had diminished over time while others, such as the *G. senegalensis*, remained the same.

Farmers have many uses for trees and shrubs. Wood is most important for rural and urban households because it is the main source of domestic energy. Based on a daily firewood consumption of 0.8 kg wood per capita, which was estimated for the population of Niamey (Projet Energie II 1991), a total of 500 000 tons of firewood was consumed in Niger in 1990. This accounted for a production deficit of 200 000 tons, subsequently, there was a net loss of trees and wood. Firewood consumption in Niger's capital, Niamey, increased in the last decade by about 3.2% per year, which is very close to the annual increase in population since 1977 (Statistisches Bundesamt 1992). *Combretum* species particularly are preferred for firewood (Fig. 4.3–6) as was also observed in other Sahelien countries such as Burkina Faso (März 1993). Good firewood burns slowly, provides a hot fire with as little smoke as possible, is easy to split, and dries quickly.

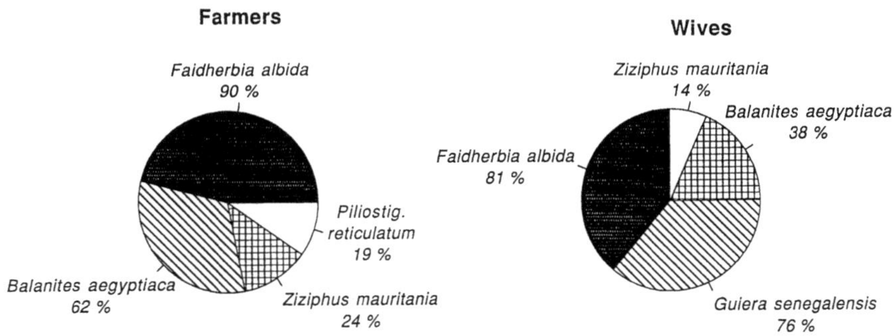

Fig. 4.3–7. Trees and bushes considered to be good browse species, according to farmers (n = 21) and their wives (n = 20) in western Niger in 1992.

Farmers also use wood from trees and shrubs for other domestic purposes: to make tools such as mortars, furniture, benches, boxes, several music instruments including drums, and working equipment such as yokes, shafts, and handles. Branches from trees and shrubs are employed as construction material for houses, huts, and store houses, while thorny branches are used for fencing against theft and livestock devastation.

The 25 *Acacia* species in West Africa represent the bulk of the wooden biomass in the region. These species are appreciated by the rural people for more than just firewood (Guinko 1992). Both farmers and their wives agreed on the importance of trees for livestock production. *F. albida* was judged by both farmers and their wives to be the most preferred species for feed (Fig. 4.3–7). The gender differences in ranking the tree species such as *B. aegyptiaca* might be related to the kind of animals kept by men and women. Men specialize in holding cattle, camels, donkeys, and horses and women prefer keeping small ruminants such as goats and sheep. *B. aegyptiaca* is, furthermore, another species that has a similar growth pattern to *F. albida* (Zech 1981).

The survey showed that farmers and their wives did not mention just one single tree that was important to them, but named at least 20 different species. Each species was valued for diverse reasons. In Niger, the value of *F. albida* can be derived from an edict of the Sultan of Damaragan, Zinder in the middle of the 19th century. He stated that "the hand would be cut off of anyone who cut a *gao*" (Haussa name for *F. albida*). Later it was changed to a saying "whoever fells a tree, cuts off his own head" (Sowers and Issouffou 1992). The existence of such laws may help explain the high density of *F. albida* trees in farmers' fields in the region of Zinder. Despite the multiple uses of trees and shrubs, little information exists on the economic value of the wooden and non-wooden parts of trees (Guinko and Pasgo 1992).

Conclusions

Although two-third of the farmers noticed that wind erosion had occurred on their own fields within the last decade, they did not consider it to be a major concern. Wind erosion damage to millet plants was mentioned by most of the farmers surveyed. Changes in topography and decreased soil fertility resulted in the formation of crusty spots (*gangani* and *tanka*). Farmers have developed a strategy to restore these spots. Millet crop residues and other mulching materials play a key role in the soil restoration process. Mulch is known to capture aeolian material and to improve the soil fertility at the same time. However, Nigerien households also use millet crop residues to feed livestock and to satisfy many domestic purposes. The survey showed that farmers' principal problem is neither awareness of the erosion problem nor availability of possible intervention techniques, but the lack of resources to transfer acquired knowledge into effective actions. The research recommendations involving crop residues seem to also have disadvantages, as they increase labor demands and augment competition between the alternative uses of crop residues (Chapter 4.2). This might obstruct farmers' actions as well.

The tree survey showed that farmers and their wives mentioned 20 different species important to them. Farmers' appreciation of tree species was strongly related to agricultural issues; they valued *F. albida* as the most preferred tree for its soil enrichment properties and as an animal feed. Their wives, however, concentrated on household needs and ranked *C. glutinosum* first. Afforestation projects are therefore well-advised to include gender in their planning and execution phases. The implementation of participatory approaches in afforestation projects have been very promising in Bangladesh (Hocking and Islam 1994) and in Burkina Faso (Ouedraogo *et al.* 1994).

The development of new agroforestry systems could gain substantially from the indigenous knowledge of rural people. Much of the knowledge about trees from the rural people surveyed has been confirmed by scientific research, even though some specific knowledge could not be verified. When socio-economic and cultural surveys are integrated into technical-oriented experiments, relevant and useful impacts for the development of sustainable farming systems can be expected. Therefore, a technological base needs to be created that blends both traditional and modern knowledge to solve the problems of farmers.

Abstract

The planning, development, and implementation phases of a wind erosion control program call for research concerning not only the technical aspects but also the indigenous knowledge about wind erosion. In 1992, two surveys were conducted among household heads and their wives in two villages in the vicinity of the ICRISAT Sahelian Center, Sadoré (Niger), to determine farmers' knowledge on wind erosion and their traditional control techniques. Farmers stated that the principle problems on their fields from wind erosion during the last decade were injuries to millet plants (69% of the 30 household heads interviewed), soil losses (45%), and surface crusting (21%). Yet, 66% of the farmers did not consider these problems a major concern since they have developed a strategy to deal with them. For instance, in the restoration of crusty spots they use organic material such as millet straw from more productive spots as a mulch. However, farmers use millet straw in many ways apart from soil erosion control. It is used to increase soil fertility, to feed livestock, to build houses or fences, and to serve other domestic purposes. The local knowledge of 21 farmers and their 20 wives was also assessed with regard to the uses of local trees and shrub species within the existing farming systems. The farmers' appreciation of tree species was strongly related to soil fertility and soil conservation management. Farmers were aware of the crop competition effects of different tree species. They valued *Faidherbia albida* as the most preferred tree, not only because of its soil enriching property but also for its value as livestock feed. Farmers' wives, however, concentrated on household needs and ranked *Combretum glutinosum* first. The two surveys showed that neither the awareness of wind erosion nor the availability of possible intervention techniques was lacking, but the resources to implement the acquired knowledge were not available. The profound knowledge of the local trees and shrubs and the importance of certain species for the local population provide a good starting point for any planning phase.

Résumé

Les savoirs paysans relatifs à l'érosion éolienne et sa maîtrise

Les étapes de planification, de développement et d'implantation d'un programme de lutte contre l'érosion éolienne nécessitent des recherches concernant non seulement les aspects techniques mais aussi les savoirs indigènes sur le sujet. En 1992, deux études ont été menées parmi les chefs de famille et leurs épouses dans deux villages à proximité du Centre Sahélien de l'ICRISAT, à Sadoré (Niger) afin de déterminer les savoirs des paysans quant à l'érosion éolienne et leurs techniques traditionnelles pour la combattre. Les paysans ont déclaré que les effets néfastes sur les plants de mil (69% des 30 chefs de famille interrogés), la perte du sol (45%) et le développement des zones de cuirasses (20%) représentaient les problèmes majeurs de l'érosion éolienne. Néanmoins, 66% des paysans ne les considéraient pas comme un problème parce qu'ils ont développé une stratégie à leur encontre. Pour réhabiliter par exemple les zones de cuirasses ils appliquent un mulch à partir de matière organique, comme de la paille de mil provenant de lieux plus productifs. Ils ont cependant trouvé beaucoup d'autres usages à la paille de mil hormis la maîtrise de l'érosion du sol tels qu'augmenter la fertilité du sol, nourrir le bétail, construire des maisons ou des haies, et autres usages domestiques. Concernant le rôle des espèces locales d'arbres et d'arbustes dans les systèmes de production existants, les savoirs des 21 paysans et de leurs 20 épouses ont également été établis. Les appréciations des fermiers pour les espèces d'arbres sont très fortement liées à leur intérêt agricole. Les paysans sont conscients des effets concurrentiels de la culture de différentes espèces d'arbres. Leur préférence va au *Faidherbia albida* pour son enrichissement du sol et son usage alimentaire du bétail. Leurs épouses cependant, plus préoccupées des besoins domestiques, classent le *Combretum glutinosum* en tête. Les deux études montrent que ni la conscience du problème ni l'existence de techniques d'intervention ne manquent, mais que seules les ressources qui permettraient la mise en oeuvre des savoirs acquis font défaut. La connaissance intime des arbres et arbustes locaux et l'importance de la spécificité des espèces pour la population locale, constitue un bon point de départ pour toute intervention dans ce domaine.

References

Angstreich, M. 1991. Are trees effective against desertification? Experiences from Niger and Mali. pp. 141–151. *In* P.T.W. Baxer (ed.) When the grass is gone: Development in intervention in African arid lands. Scandinavian Institute of African Studies, Uppsala, Sweden.

Baidu-Forson, J. 1988. Characteristics of farm households and economy in western Niger villages: Some evidence from baseline surveys. Resource Management Program, Economics Group, Progress Report 87, International Crops Research Institute for the Semi-Arid Tropics, Niamey, Niger.

Bonnard, P. and S. Scherr. 1994. Within gender differences in tree management: Is gender distinction a reliable concept? *Agroforestry Systems* 25: 71–93.

Brouwer, J., L.K. Fussell, and L. Herrmann. 1993. Soil and crop growth micro-variability in the West African semi-arid Tropics: A possible risk-reducing factor for subsistence farmers. *Agriculture, Ecosystems and Environment* 45: 229–238.

Buerkert, A., R.D. Stern, and H. Marschner. 1995. Post stratification clarifies treatment effects on millet growth in the Sahel. *Agronomy Journal* 87: 752–761.

Chase, R. and E. Boudouresque. 1987. Methods to stimulate plant regrowth on bare Sahelian forest soils in the region of Niamey, Niger. *Agriculture, Ecosystems and Environment* 18: 211–221.

Feil, P.R., J.P.A. Lamers, and L. Herrmann. 1995. Knowledge transfer in the field: Solving crop residue problems in Niger. *Journal of Technology Transfer* 20(1): 31–41.

Gavian, S. 1992. Manure and crop residue management strategies used by farmers in selected regions of the Tillabéri department, Niger. ILCA Niger Research Report 92/1, Niamey, Niger.

Geiger, S.C. and A. Manu. 1993. Soil surface characteristics and variability in the growth of millet in the plateau and valley region of western Niger. *Agriculture, Ecosystems and Environment* 45: 203–211.

Geiger, S.C., A. Manu, and A. Bationo. 1992. Changes in a sandy Sahelian soil following crop residue and fertilizer additions. *Soil Science Society of America Journal* 56: 172–177.

Gross, G. 1987. A numerical study of air flow within and around a single tree. *Boundary-Layer Meteorology* 15: 81–93.

Guinko, S. 1992. Rôle des acacias dans le développement rural au Burkina Faso et au Niger, Afrique de l'Ouest. pp. 3–16. *In* R. Wittig and S. Guinko (eds.) Etudes sur la flore et la végétation du Burkina Faso et des pays avoisinants. Vol. I. Verlag Natur & Wissenschaft, Solingen, Germany.

Guinko, S. and L.P. Pasgo. 1992. Récolte et commercialisation des produits non ligneux des essences forestières locales dans le département de Zitenga, province d'Oubritenga, Burkina Faso. pp. 17–25. *In* R. Witti and S. Guinko (eds.) Etudes sur la flore et la végétation du Burkina Faso et des pays avoisinants. Vol. I. Verlag Natur & Wissenschaft, Solingen, Germany.

Hailu, Z. and A. Runge-Metzger. 1992. Sustainability of land-use systems: The potential of indigenous measures for the maintenance of soil productivity in sub-Sahara African agriculture. Tropical Agroecology 7, Margraf Verlag, Weikersheim, Germany.

Hocking, D. and K. Islam, 1994. Trees in Bangladesh paddy fields and homesteads: Participatory action research towards a model design. *Agroforestry Systems* 25: 193–216.

Hoogmoed, W.B. and L. Stroosnijder. 1984. Crust formation on sandy soils in the Sahel. I. Rainfall and infiltration. *Soil and Tillage Research* 4: 5–23.

ICRISAT. 1992. ICRISAT West African Programme, Annual Report 1991. ICRISAT Niamey, Niger.

Kainkwa, R.M.R. and C.J. Stigter. 1994. Wind reduction downwind from a savanna woodland edge. *Netherlands Journal of Agricultural Science* 42(2): 145–157.

Kerkhof, P. 1990. Agroforestry in Africa: A survey of project experience. pp. 113–123. *In* G. Foley and G. Barnard (eds.) Panos Technical Series. Panos Publications London Ltd.

Lamers, J.P.A., K. Michels, B.E. Allison, R.J. Vandenbeldt, and M. von Oppen. 1994a. Trees for agroforestry systems in Niger: Researchers' gains through local knowledge of men and women. *Quarterly Journal of International Agriculture* 33(2): 179–191.

Lamers, J.P.A., K. Michels, and R.J. Vandenbeldt. 1994b. Trees and windbreaks in the Sahel: Establishment, growth, nutritive and calorific values. *Agroforestry Systems* 26(3): 171–184.

Long, S.P. and N. Persaud. 1988. Influence of neem (*Azadirachta indica*) windbreaks on millet yield, microclimate, and water use in Niger, West Africa. *In* Proceedings International Conference on Dryland Farming, Bushland/Amarillo, TX, 19–22 Aug. 1988.

März, U. 1993. The determination of mixing ratios in afforestation assortments: Satisfying farmers' needs for multi-purpose forests. *Journal of Agriculture in the Tropics and the Subtropics* 94: 175–183.

Michels, K. 1994. Wind erosion in the southern Sahelian zone. Diss. Universität Hohenheim, Verlag Ulrich E. Grauer, Stuttgart, Germany.

Powell, M. and F. Ikpe. 1992. Fertilizer factories: Nutrient recycling through livestock. *ILEIA Newsletter* 8(3): 13–14.

Projet Energie II – Energie Domestique. 1991. Volet Offre. Schema Directeur d'Approvisionnement en Bois Energie de Niamey. Groupement SEED-CTFT, République du Niger, Ministère de l'Hydraulique et de l'Environnement, Niamey, Niger.

Ouedraogo, K., O. Traoré, and K. Nebie. 1994. Forest management and farmers' participation. The case of Bougnouou village in the province of Sissili, Burkina Faso. *Forest, Trees and People Newsletter* 24: 44–47.

Rebafka, F.P. 1993. Deficiency of phosphorus and molybdenum as major growth limiting factors of pearl millet and groundnut on an acid sandy soil in Niger, West Africa. Hohenheimer Bodenkundliche Hefte 9, Stuttgart, Germany.

Renard, C. and R.J. Vanderbeldt. 1990. Bordures d'*Andropogon gayanus* Kunth. comme moyen de lutte contre l'érosion éolienne au Sahel. *Agronomie Tropicale* 45(3): 227–231.

Sall, P.N. 1992. Site effects of *Acacia albida* Del. pp. 103–105. *In* R.J. Vandenbeldt (ed.) *Faidherbia albida* in West African semi-arid tropics. Proceedings Workshop, ICRISAT and ICRAF, Niamey, Niger, 22–26 April 1991, ICRISAT, Patancheru, India and ICRAF, Nairobi, Kenya.

Schulze, E.D., G. Gebauer, H. Ziegler, and O.L. Lange. 1991. Estimates of nitrogen fixation by trees on an aridity gradient in Namibia. *Oecologia* 88: 451–455.

Shaxson, T.F. 1985. Erosion, economics, subsistence and psychology. pp. 667–673. *In* S.A. El-Swaify, W.C. Moldenhauer, and A. Lo (eds.) Soil erosion and conservation. Soil Conservation Service, Aukeny, IA.

Sowers, F.W. and M. Issouffou. 1992. Precolonial agroforestry and the implications for the present: The case of the Sultanate of Damagaram, Niger. pp. 171–175. *In* R.J. Vandenbeldt (ed.) *Faidherbia albida* in West African semi-arid tropics. Proceedings Workshop, ICRISAT and ICRAF, Niamey, Niger, 22–26 April 1991, ICRISAT, Patancheru, India and ICRAF, Nairobi, Kenya.

Speirs, M. and O. Olson. 1992. Indigenous integrated farming systems in the Sahel. World Bank Technical Paper 179, Africa Technical Department, Washington D.C.

Statistisches Bundesamt. 1992. Länderbericht Niger 1992. Verlag J.B. Metzler und C.E. Poeschel, Wiesbaden, Germany.

Taylor-Powell, E., A. Manu, S. Geiger, M. Ouattara, and A.S.R. Juo. 1991. Integrated management of agricultural watersheds: Land tenure and indigenous knowledge of soil and crop management. TropSoils Bulletin No. 91–04, Soil Management Collaborative Research Support Program, North Carolina State University, Raleigh, NC.

Vandenbeldt, R.J. and J.H. Williams. 1992. The effect of soil surface temperatures on the growth of millet in relation to the effect of *Faidherbia albida* trees. *Agricultural and Forest Meteorology* 60: 93–100.

Zech, W. 1981. Futterwert forstlicher Pflanzen im Sahel. Forschungsberichte des Bundesministeriums für Wirtschaftliche Zusammenarbeit. Vol. 7. Weltforum-Verlag, München, Köln, London.

PART FIVE

Summary and Outlook

Summary and Outlook

Matthias von Oppen and John P.A. Lamers

Background

The Sahara is the world's largest source of dust which is stirred and carried away by repeated seasonal convective storms. Within the Sahara, winds transport massive amounts of airborne dust and saltated soil. Subsequently, this dynamic transport process results in the redeposition of dust and soil throughout the region. Towards the fringe areas of the Sahara such as in the Sahel, deposition and erosion of soil and dust alternate during the year. The extent of erosion depends not only upon the season and weather conditions, but also upon human intervention.

Convective storms add to the vagaries of climatic risk in subsistence agricultural production in the Sahel. Soil erosion may deplete specific fields of valuable topsoil and impede crop production. Sandstorms may also destroy young plants by covering them with wind-blown soil or cause abrasive damage through sand blasting. However, where wind-blown soil and dust are deposited in a field, new soil layers may be formed and nutrients may be added to the poor soils in the region.

Researchers faced with the task of understanding the situation and deciding about possible alternatives for improvement approached this task in steps. In view of the often contradictory effects of wind erosion on overall productivity in the Sahel, it was logical to first ask questions about the extent and the implications of wind erosion and dust transport. Not only damage to soils and crops have to be considered but also possible benefits (Tables 5–1 and 5–2). This then led to experiments on suitable measures to reduce the extent and the effects of wind erosion and these provided the basis for recommending interventions. Economic assessment helped define the cost-effective interventions which aimed at minimizing the negative effects of wind erosion and/or maximizing the positive effects of wind control. The assessment also took into consideration that both individual farmers and society as a whole have a common interest in reducing erosion and soil degradation.

B. Buerkert, B.E. Allison and M. von Oppen (eds.), Wind Erosion in Niger, 235–244.

Table 5–1. Effects of measures to control wind erosion on crop growth, wind speed, and soil flux

Control measures	Year *Chapter*	Effects on crop growth and yield	Effects on wind speed and soil flux
Soil ridging	1985–87 *3.1*	• small effects; in cowpea greater than in millet	• effects negligible
Crop residue mulch at 2000 kg CR ha^{-1}	1992–94 *3.3, 3.4 4.2, 4.3*	• large increase in millet grain and straw yields • increase in plant stand • more weeds	• reduction in soil flux • increase in soil depth by deposition and reduced erosion (*3.3*)
Crop residue mulch at 500 kg CR ha^{-1}	1992–93 *3.4, 4.2 4.3*	• increase in millet grain and straw yields, only in 1993 • small effects on plant stand	• no effects on soil flux
Natural savannah windbreaks with spacings of 6, 20, 40, and 90 m	1985–87 *3.1*	• larger increase in early dry matter production of millet (6 m, 20 m spacings) • large increase in early ground cover of cowpea (6 m, 20 m spacings) • small effects on grain and straw yields of cowpea and millet	• reduction in wind speed and wind-blown soil (6 m, 20 m spacing only) because of low height)
Planted wind-break tree and grass species	1988, 1992–93 *3.2, 4.1, 4.3*	• no effects on plant stand • increase in millet yields only with *F. albida* windbreaks • reduction in millet yields in the vicinity of windbreaks	• large reduction in soil flux with *B. rufescens* windbreaks; effects smaller with *A. gayanus* • 14 m area protected with *B. rufescens*; 10 m area protected with *A. gayanus*

Extent of Damage to Plants and Soils

In western Niger, an average of 10 to 20 convective storms, lasting 15 to 60 minutes, occur at irregular intervals during the beginning of the growing season. If such storms occur where millet (*Pennisetum glaucum* (L.) R. Br.) seedlings are just emerging or are a few days old, they may cause damage by covering and, thereby, retarding growth or destroying the plants. This may cause partial or total crop failure forcing farmers to replant the field. If sandstorms occur at later growth stages they can cause abrasive damage to the stems and leaves by sand blasting; however, millet plants are capable of recovering from such damage. Thus, pearl millet production in Niger may be affected by wind erosion only if severe sandstorms occur during the early sensitive stage of millet establishment when emerging seedlings are covered

Table 5–2. Effects of measures to control wind erosion on nutrient and water supply, and financial returns and costs

Control measures	Year *Chapter*	Effects on nutrient and water supply	Financial returns and costs
Soil ridging	1985–87 *3.1*	• not determined	• time and labor required for ridging must be considered
Crop residue mulch at 2000 kg CR ha^{-1}	1992–94 *3.3, 3.4* *4.2, 4.3*	• increase in cation exchange capacity in the topsoil • high nutrient supply to the crop • increase in organic carbon (*3.4*)	• high amounts of CR are required, thus, production of CR must be increased • competition with other CR uses • increase in weeding time • cash availability to hire labor restricts the CR application • weeds for folder
Crop residue mulch at 500 kg CR ha^{-1}	1992–93 *3.4, 4.2* *4.3*	• no effects on cation exchange capacity • no effect on organic carbon	
Natural savannah windbreaks with spacings of 6, 20, 40, and 90 m	1985–87 *3.1*	• subsoil water content lower at the windbreak crop interface • subsoil water content lower on plots with the 6 m spacing during periods of low rainfall • competition between crops and windbreak vegetation for water and nutrients is probable	• low cost for establishment and maintenance and low labor requirements are expected • additional fodder from windbreak vegetation • large areas of land required for the windbreak strips
Planted wind-break tree and grass species	1988, 1992–93 *3.2, 4.1* *4.3*	• increase in cation exchange capacity, organic carbon, and nitrogen in the topsoil, but only in the vicinity of the windbreaks • competition between crops and windbreak vegetation for water, nutrients, and light is probable	• additional costs and labor for establishment and maintenance • additional commodities: firewood, timber, fruits, fodder • *F. albida* and *A. holosericea* had highest returns when only millet grain yield was considered • *A. senegal* and *A. indica* are good for investments in firewood • *A. senegal* and *A. nilotica* are good for investments in feed • returns to family labor for *A. gayanus* and *B. rufescens* were lower than the control with no windbreaks

with wind-blown soil or completely uprooted by the sandstorm. After that period, the short sandstorms are not likely to cause significant wind erosion damage to pearl millet.

During convective storms most wind-blown soil is transported near the soil surface (0.1 m height above ground level), indicating a rather short-distance transport compared to dust which is transported over long distances. Wind erosion can cause a loss of 140 t ha^{-1} of fertile topsoil within two months on unprotected sandy soils. At the same time, similar amounts of eroded soil can be deposited in areas where wind barriers or a surface mulch of 2000 kg ha^{-1} of crop residues trap the transported soil. This implies a redistribution of soil particles and, consequently, nutrients which on individual farms may increase or decrease the soil fertility. The consequence is an increase in the variability of plant growth which is typical for farmers' fields in the West African Sahel.

Control Measures for Wind Erosion

Even though the damage caused by sandstorms was somewhat less dramatic than initially expected, damage to soils and plants does occur. Furthermore, there are positive effects on plant growth and on biomass production associated with reducing the wind speed because of microclimatic reasons. Therefore, alternative types of wind erosion control and their effects were studied. The difficulty in assessing such effects is that various compounding factors must initially be investigated separately. This was attempted in the course of several studies. The alternatives studied for reducing wind erosion ranged from (i) roughening the soil surface through ridging, (ii) applying crop residues in various amounts, (iii) leaving stripes of natural savannah growth to form windbreaks at various distances, and (iv) establishing trees in rows as windbreaks (Table 5–1).

(i) Ridging is an effective measure to control wind erosion since it reduces soil fluxes, but the effect on yields is not significant. If at all, cowpea (*Vigna unguiculata* (L.) Walp.) plants respond better than millet with slightly higher yields in ridged compared to flat fields.

(ii) Crop residues were applied as mulch at the rates of 500 and 2000 kg ha^{-1}. A mulch of crop residues at a rate of 2000 kg ha^{-1} reduces wind speed and, consequently, soil flux. However, a mulch at a rate of 500 kg ha^{-1} has no effect on wind erosion. The critical level of crop residues necessary to reduce erosion appears to be between 800 and 2000 kg ha^{-1}, however it is not yet known precisely. Only the higher rate of 2000 kg ha^{-1} of crop residues has significant effects on crop yields. This is not only due to a significant reduction in soil flux and additional soil deposition between the crop residues, but also due to a supply of nutrients from the mulch. However, the amount of crop residues harvested in a normal year is less than 2000 kg ha^{-1}, making such a rate of application impractical. Also, considering the value of millet crop residues as fodder and for many domestic uses, its application by farmers to reduce wind erosion may be restricted to specific areas in a field where damage due to wind erosion is particularly severe. Concentrating available crop residues on areas with a low productivity is a short-term strategy.

(iii) Leaving the natural savannah vegetation in strips perpendicular to the main wind direction and cultivating between them, generates a system of windbreaks that protects the crop at relatively low costs. Research shows that during the three years of research, a wind erosion control effect was determined only between strips spaced 6 and 20 m apart. Given the low heights of the examined windbreaks (0.6 m) in the first year, protection at wider spacings was not possible during the experiment. Growth of millet and cowpea is positively affected, but only during early stages of development. At later growth stages, the growth advantage disappears and may even turn into a disadvantage, probably because of competition for water and nutrients between the crop and narrowly spaced windbreaks.

(iv) Trees and *Andropogon gayanus* grass planted in rows as windbreaks have measurable effects on wind erosion. The windbreaks reduce soil flux with the effects being dependent upon the windbreak species. After five years, a measurable change in soil properties is found only in close proximity to the windbreaks. The results of these experiments show that at a distance from the windbreak up to 10 times its height, millet growth responds positively to windbreaks. Moreover, certain species such as *Faidherbia albida* increase millet grain and dry matter yields. However, the tree windbreaks may not justify their costs unless the value of by-products, in terms of wood and fodder, is high enough.

Economic Analyses and Farmers' Views on Alternative Techniques

The assessment of alternative methods of reducing wind erosion and improving crop production shows that crop residues and tree windbreaks have some promising potential in technical terms. In order to further assess the potential, an evaluation in terms of their financial and economic performance was undertaken. Also, a farm survey assessing farmers' opinions on these options was carried out. The economic assessment was based on direct (tree product) and indirect (protection by windbreaks) benefits. The results show that the economic value of the wood and non-wood parts of trees, as well as other plants, considerably affects returns to windbreaks. Therefore, it is important to include these features in an economic analysis. In the present analysis, a lack of information prevented the inclusion of some benefits, such as from pods of *F. albida* as feed, gum arabic of *Acacia senegal*, bark of *Acacia nilotica* for tanning leather, timber of *Azadirachta indica*, the straw of *A. gayanus*, and the medicinal properties of all species. In addition, when anticipating windbreak planting, a diversity of species should be advocated. In the context of a farm household, crop residue applications show promising effects, even though labor is a limiting factor. Farmers' assessment of the trees and crop residue applications confirmed the results of the financial and economic analyses performed in this study. The picture that emerged from these evaluations was that: (i) among the trees, *F. albida* is the most preferable one to use as a windbreak, even if not planted in rows and (ii) the application of crop residues should be concentrated in areas of low productivity within the fields. However, since it is likely that in the future land will become more and more scarce, mulching to control wind erosion will be competitive with traditional crop residue uses.

Outlook and Recommendations

Future Research Needs

Much of the on-station research conducted so far in the West African Sahel has produced excellent techniques to reduce soil degradation and increase soil fertility. However, these techniques have not been widely accepted in local farming systems. Sahelian households have many goals and undertake a variety of activities to achieve these goals. Agriculture, livestock, commerce, art, and off-season labor abroad are some of the activities which are dispersed over the year and which may have long, medium, or short-term perspectives. The many economic interests of rural households leave very little labor available for the implementation of innovations, particularly for controlling wind erosion. Thus, every recommended technology must be flexible enough to match the diversity and variability of the farming systems in the region. In order to ensure this, socio-economic and socio-cultural surveys should be integrated into technology-oriented experiments, thus calling for an interdisciplinary approach. Furthermore, the systematic participation of farmers in assessing and adapting innovations should become a general practice.

Analytical tools for a long-term analysis are still not sufficiently developed and need further research. These tools should incorporate the dimension of time, consider the dynamics of factors like wind erosion in farming systems, and extrapolate farm-level results to a larger scale, be it farm, region, or nation, as well as quantify and value environmental damage and improvement. Also, in order to account for the factors of risk and uncertainty, both of which influence the decisions of farmers and the adoption of erosion control techniques, long-term data quantifying the relationship between erosion control with windbreaks (or, in fact, any erosion control technique) and the financial returns over time, are needed.

Social scientists are calling for the development of analytical tools in the fields of soil science and economics to evaluate the different levels of land erosion, because decisions on allocating limited resources for erosion control cannot be based solely on benefit/cost analyses. Socially relevant aspects such as the sustainability of systems, regional parity, and other equity effects influenced by soil erosion have not been considered from a farmer's perspective. Social analysis should, therefore, be included when decisions on a national or even regional level are to be made. The choice of a social discount rate, instead of the commercial discount rate for capital, poses particular difficulties in the context of assessing such long-term phenomena as soil erosion.

In order to focus on the possible implementation of different soil erosion control techniques, future research should also be directed towards regional surveys and characterization of the areas affected by wind erosion. Important tools for this purpose are low-cost aerial photography (Buerkert *et al.* 1995) or satellite images to determine land-use and vegetation, as well as the use of geographic information systems to improve the interpretation of such information (Quiroz *et al.* 1995). A valuable outcome of such research would be a region-wide wind erosion probability map, with reference to the ideal and adapted techniques of erosion control measures and in association with the different crops important for the region.

Since pearl millet is the main staple in this region, most research has been conducted on this crop. However, increased millet production by itself may not be sufficient to justify expensive innovations because millet generally has a low market value. More research is required on the response of other crops with higher economic value such as cowpea and groundnut.

Policy and Institutional Recommendations

If erosion control measures based on tree planting are to be accepted by the rural people, they should be simple; based on indigenous knowledge; inexpensive with no demands for nurseries, special tools, or supplementary watering methods; profitable in the short-term; and manageable by individual households in order to minimize community organization. Even then, such ideal solutions still demand institutional and political support for implementation and the cost of this support should not be underestimated.

An important criterion for the success of implementing soil erosion control techniques is the full acceptance of the responsibility for the management of natural resources by the rural population, as they are the beneficiaries. Literature on land tenure points out that due to the insecurity of tenancy status, tenants are reluctant to invest in soil conservation practices and particularly in investments with long pay-back periods such as windbreaks (Kerkhof 1990; Neef *et al.* 1996). Land tenure, ownership of land and trees, and other property rights should be clarified before erosion control activities are undertaken on a large scale. In this context, individual ownership may not necessarily be the only answer. Awareness of the need for environmental protection and soil conservation as well as individual motivation to undertake such action can also be developed for commonly owned properties. This awareness and motivation is essential for a successful implementation of any erosion control technique. Motivating farmers to use innovations which primarily have a long-term profitability is vital because it would allow them to sustain their families and maintain their resource base. The motivation of farmers is a core task of extension, but policy and institutional support must also be assured. This requires motivation of the community as a whole, and decision-making by the community should include representatives of all interested groups in the society.

Farmers should be incorporated not only during the execution and maintenance of tree planting, but already at the identification and planning phases. The program put in place should address not only individual fields but also the management of village land. Extension staff should be encouraged to play the role of a moderator by guiding decision-making within the community but should avoid giving an opinion on particular issues or influencing decision-making. The time needed to convince and mobilize communities should not be underestimated as it can take several years.

Given the importance of extension programs, the individual extension organization needs to be considered when defining extension activities for windbreaks. The planting of trees and windbreaks is normally a responsibility of the forestry department. However, in many Sahelian countries these departments are structured as a para-military organization, having an extension function and a police and control function at the same time (Lamers 1996). Because of the dual and contradictory nature of their assignment, the success of these agencies is minimal. Policy and institutional objectives of most local governments need to be re-evaluated if a successful implementation of any soil conservation technique is to be achieved.

Soils prone to erosion will lose valuable soil nutrients. If the protected soil is cultivated, yield increases are anticipated through the protection of millet by windbreaks, but this increases the export of nutrients from the field in harvested products as well. If these nutrients are not replaced, soil depletion will occur faster with windbreaks than without windbreaks. From an agronomic viewpoint, the use of organic and mineral fertilizers is imperative. Sahelian farmers often apply them at sub-optimal rates due to the limited availability of organic fertilizers, high prices, and the inaccessibility of mineral fertilizers; or they do not use mineral fertilizers at all because of the unreliable rains or a lack of capital. The accessibility and distribution of inexpensive inputs must, therefore, be assured and demands an immediate intervention from policy-makers. Without the use of mineral fertilizers, it is unlikely that erosion in the Sahel will be reduced.

Soil erosion control is a task which goes far beyond a farmer's and even a Sahelian nation's level. Without stopping erosion now, the Sahelian population, who up to this point have managed to survive, will in the future require food support from external sources at excessively high costs. This support would, in all likelihood, far exceed the present costs of implementing soil conservation technologies. Postponing the implementation of soil conservation techniques is questionable because the opportunity costs of farmers' foregone income are high. Worldwide, wind erosion control techniques have not been implemented on a large scale without financial and economic incentives from farm-external sources (Lamers *et al.* 1996). Therefore, it is unrealistic to expect that Sahelian farmers will implement wind erosion control measures on their own without financial support. Also, economic incentives and favorable agricultural price policies are required to reinforce the impact of research and extension. Establishing more markets for firewood and feed may assist in the adoption of planting trees, either as rows of windbreaks or as individual trees in a scattered park landscape.

Regarding the immense activities within and around the Sahara dust bowl, the impact of individual human interventions to stabilize soil fertility may indeed seem hopelessly small. However, developing sustainable production systems in the Sahel is crucial. If the task of maintaining soil fertility is accepted as an international responsibility, ways could be found to implement measures of erosion control. The techniques which have been described in this book may not be very powerful for an overall solution of the problem of wind erosion. However, these techniques are available and can be effective under certain conditions, thus, the challenge remains to move forward with their judicious implementation based upon individual as well as national and international support.

References

Buerkert, A., F. Mahler, and H. Marschner. 1996. Soil productivity management and plant growth in the Sahel: Potential of an aerial monitoring technique. *Plant and Soil* 180(1): 29–38.

Kerkhof, P. 1990. Agroforestry in Africa: A survey of project experience. pp. 113–123. *In* G. Foley and G. Barnard (eds.) Panos Technical Series. Panos Publications London Ltd.

Lamers, J.P.A. 1995. An assessment of wind erosion control techniques in Niger: Financial and economic analyses of windbreaks and millet crop residues. Diss. Universität Hohenheim, Verlag Ulrich E. Grauer, Stuttgart, Germany.

Lamers, J.P.A., T. Becker, and M. von Oppen. 1996. Arrangements to finance tree windbreaks for wind erosion control. pp. 227–241. *In* B. Buerkert, B.E. Allison, and M. von Oppen (eds.) Wind erosion in West Africa: The problem and its control. Proceedings International Symposium, Hohenheim, Germany, 5–7 Dec. 1994, Margraf Verlag, Weikersheim, Germany.

Neef A., J. Haigis, and F. Heidhues. 1996. L'impact de pluralisme institutionnel et juridique sur l'indroduction de mesures anti-érosives – le cas du Niger. pp. 283–296. *In* B. Buerkert, B.E. Allison, and M. von Oppen (eds.) Wind erosion in West Africa: The problem and its control. Proceedings International Symposium, Hohenheim, Germany, 5–7 Dec. 1994, Margraf Verlag, Weikersheim, Germany.

Quiroz, R.A., R.B. Estrada, C.L. Velarde, and H.G. Zandstra. 1995. The use of simulation models in developing sustainable management of natural resources for the Andean zone. *International Potato Center (CIP) Circular* 21(1): 10–18.

Synthèse et perspectives

Matthias von Oppen et John P.A. Lamers

Contexte

Le Sahara est la plus grande source de poussière du monde, soulevée et emportée lors d'orages de convection saisonniers récurrents. Ces vents transportent dans l'espace saharien des masses importantes de poussières aérosols et de sols salés. De ce processus dynamique de transport résulte une redéposition de poussière et de sol dans toute la zone. Vers les aires limitrophes du Sahara comme le Sahel, érosion et dépôt de sol et de poussière alternent tout au long de l'année. L'ampleur de l'érosion dépend de la saison et des conditions climatiques mais aussi de l'intervention humaine.

Les orages de convection ajoutent aux caprices du risque climatique qui conditionne la production agricole de subsistance dans le Sahel. L'érosion du sol peut épuiser des zones déterminées de terre arable fertile et faire obstacle à la production agricole. En outre les tempêtes de sable peuvent détruire les jeunes plants en les recouvrant de sol transporté par le vent ou causer des dégâts dûs à l'effet abrasif des rafales de sable. Toutefois, sur les parcelles où les poussières aérosols se déposent, de nouvelles couches de sol peuvent se constituer et des éléments nutritifs peuvent s'ajouter aux sols appauvris de la région.

Les chercheurs confrontés à la tâche de comprendre la situation et de choisir parmi des alternatives possibles d'amélioration ont procédé par étapes. En regard des effets souvent contradictoires de l'érosion éolienne sur la productivité globale dans le Sahel, la logique revenait à tout d'abord s'interroger sur l'ampleur et les effets de l'érosion éolienne et des transports de poussière. On devait prendre en compte les dommages causés aux sols et aux cultures mais aussi les éventuels effets bénéfiques (tableaux 5–1 et 5–2). Ceci a débouché sur des expérimentations de mesures appropriées visant à réduire l'extension et l'impact de l'érosion éolienne, qui ont fourni la base pour des recommandations quant aux interventions. L'évaluation économique a aidé à définir les interventions à coût réel destinées à réduire les effets négatifs de l'érosion éolienne et/ou à maximiser les effets positifs de la maîtrise du vent.

B. Buerkert, B.E. Allison and M. von Oppen (eds.), Wind Erosion in Niger, 245–255.

L'évaluation a aussi pris en considération le fait que les exploitants individuels mais aussi la société dans son ensemble avaient un intérêt commun à la diminution de l'érosion et de la dégradation des sols.

Ampleur des dégâts sur les cultures et les sols

Dans l'ouest du Niger, 10 à 20 orages de convection en moyenne dont la durée va de 15 à 60 minutes surviennent à intervalles réguliers durant le début de la période de croissance. Si de telles tempêtes touchent un champ où les plantules de mil émergent à peine ou n'ont que quelques jours, elles peuvent causer des dommages en les recouvrant retardant ainsi leur croissance ou bien en les détruisant. Cela peut affecter tout ou partie de la culture, obligeant ainsi les agriculteurs à replanter leur champ. Si les tempêtes de sable touchent les plants à des stades ultérieurs de croissance, elles peuvent avoir un effet abrasif sur les tiges et les feuilles causé par les rafales de sable; cependant, les plants de mil peuvent redémarrer après de tels dommages. C'est pourquoi la production de petit mil au Niger ne peut être affectée par l'érosion éolienne que si de graves tempêtes de sable se produisent durant la période fragile du début de sa plantation, recouvrant les plantules de sol transporté par le vent ou les déracinant complètement. Après cette période, il est peu probable que les tempêtes de sable de courte durée aient un effet érosif important sur le mil.

Pendant les orages de convection la plupart du sable transporté par le vent voyage près de la surface du sol (à 0,1 m au dessus du niveau du sol), ce qui implique un transport de plutôt courte distance en comparaison avec la poussière qui se trouve transportée sur de longues distances. L'érosion éolienne peut provoquer en deux mois sur un sol sableux non protégé, une perte de 140 t ha^{-1} de sol fertile de surface. En même temps, des quantités similaires de sol érodé peuvent être déposées à des endroits où des barrages contre le vent ou un paillage de surface de 2000 kg ha^{-1} de résidus de récoltes peuvent piéger le sol transporté. Cela implique une redistribution des particules du sol et donc des éléments nutritifs qui sur les exploitations individuelles peut augmenter ou diminuer la fertilité d'une terre. La conséquence en est une variabilité accrûe de la croissance des plantes, typique en ce qui concerne les champs des agriculteurs du Sahel de l'Afrique de l'Ouest.

Techniques de maîtrise de l'érosion éolienne

Bien que moins dramatiques qu'initialement prévu, les dommages causés aux plants de petit mil et aux sols par les tempêtes de sable sont réels. Qui plus est, il y a des effets positifs sur la croissance des plantes et sur la production de biomasse associés à la réduction de la vitesse du vent pour des raisons microclimatiques. C'est pourquoi on a étudié différents types d'érosion éolienne et

Tableau 5–1. Effets de techniques de lutte contre l'érosion éolienne sur la croissance des cultures, la vitesse du vent et les flux du sol

Techniques	Année Chapitre	Effets sur la croissance des culltures et leur rendements	Effects sur la vitesse du vent et les flux du sol
Labour du sol	1985–87 *3.1*	• peu d'effets; plus importants sur le niébé que sur le petit mil	• effets négligeables
Paillage à base de résidus de récoltes (2000 kg CR ha^{-1})	1992–94 *3.3, 3.4* *4.2, 4.3*	• augmentation importante des rendements en grains de petit mil et en paille • augmentation en pieds • plus de graines	• réduction des flux du sol • augmentation de la profondeur du sol due à la déposition et à la réduction de l'érosion (*3.3*)
Paillage à base de résidus de récoltes (500 kg CR ha^{-1})	1992–93 *3.4, 4.2* *4.3*	• augmentation des rendements en grains de petit mil et en paille, seulement en 1993 • peu d'effets sur les pieds	• pas d'effets sur les flux du sol
Brise-vent naturels de savane espacés de 6, 20, 40 et 90 m	1985–87 *3.1*	• augmentation importante de la production première de matière sèche de petit mil (espacements 6 m, 20 m) • augmentation importante de la couverture du sol en niébé (espacements 6 m, 20 m) • peu d'effets sur les rendements en grains et paille de petit mil et de niébé	• réduction de la vitesse des vents et du sol transporté par le vent • (espacements 6 m, 20 m seulement en raison de la hauteur basse)
Plantation de brise-vent (arbres et herbes)	1988, 1992–93 *3.2, 4.1,* *4.3*	• pas d'effets sur les pieds • augmentation des rendements du petit mil seulement avec des brise-vent de *F. albida* • diminution des rendements de petit mil à proximité des brise-vent	• réduction importante des flux du sol avec des brise-vent de *B. rufescens*; effet moins important avec *A. gayanus* • l'aire protégée a été de 14 m avec *B. rufescens* et 10 m avec *A. gayanus*

leurs effets. La difficulté à évaluer ces effets tient à ce que l'on doit d'abord étudier séparément divers facteurs les composant. C'est ce qui a été tenté à travers plusieurs recherches. Les possibilités étudiées pour réduire l'érosion

Tableau 5–2. Effets de techniques de lutte contre l'érosion éolienne sur les réserves en éléments nutritifs et eau et les rémunérations

Techniques	Année *Chapitre*	Effets sur l'apport en éléments nutritifs et eau	Rémunérations financières et coûts
Labour du sol	1985–87 *3.1*	• non déterminé	• le temps et le travail requis pour le labour doivent être pris en compte
Paillage à base de résidus de récoltes (2000 kg CR ha^{-1})	1992–94 *3.3, 3.4* *4.2, 4.3*	• augmentation de la capacité d'échanges cationiques du sol de surface • apport important d'éléments nutritifs à la culture • augmentation du carbone organique (*3.4* seulement)	• de grandes quantités de résidus de récoltes sont requis donc leur production doit augmenter • compétition avec d'autres usages des résidus de récoltes • augmentation du temps de sarclage • la nécessité de liquidités pour louer la force de travail limite l'application de CR • herbes pour le fourrage
Paillage à base de résidus de récoltes (500 kg CR ha^{-1})	1992–93 *3.4, 4.2, 4.3*	• pas d'effets sur la capacité d'échanges cationiques • pas d'effets sur le carbone organique	
Brise-vent naturels de savane espacés de 6, 20, 40 et 90 m	1985–87 *3.1*	• réserve en eau du sol moindre à l'interface culture brise-vent • réserve en eau du sol moindre sur les parcelles à espacements de 6 m pendant les périodes de faibles précipitations • possible compétition entre les cultures et la végétation de brise-vent pour l'eau et les éléments nutritifs	• faible coût possible pour la mise en place et l'entretien et nécessité peu de force de travail • fourrage supplémentaire tiré de la végétation des brise-vent • de grandes étendues de terrain sont nécessaires à la plantation des brise-vent en rangées

éolienne ont été classées comme suit: (i) rendre la surface du sol rugueuse par le labour, (ii) appliquer des résidus de récoltes en quantités diverses, (iii) laisser des bandes de savane naturelle pousser pour former des brise-vent à des distances variables, et (iv) planter des rangées d'arbres comme brise-vent (tableau 5–1).

Tableau 5–2. Continué

Techniques	Année *Chapitre*	Effets sur l'apport en éléments et leur rendements	Rémunérations financières et coûts
Plantation de brise-vent (arbres et herbes)	1988, 1992–93 *3.2, 4.1, 4.3*	• augmentation de la capacité des échanger cationiques, du carbone organique et de l'azote dans le sol de surface seulement à proximité des brise-vent • possible compétition entre les cultures et la végétation de brise-vent pour l'eau, les éléments nutritifs et la lumière	• coûts et travail supplémentaires pour la mise en place et l'entretien • produits supplémentaires: bois de feu, bois d'oeuvre, fruits et fourrage • *F. albida* et *A. holosericea* ont donné les plus grandes retombées si l'on tient compte du petit mil seulement • *A. senegal* et *A. indica*: bons investissements en bois de feu • *A senegal* et *A. nilotica*: bons investissements en alimentation du bétail • les rémunérations de la force de travail familial pour *A. gayanus* et *B. rufescens* étaient inférieures que sur les parcelles-témoins sans brise-vent

(i) Le labour est une mesure efficace de lutte contre l'érosion éolienne puisqu'il réduit les flux du sol, mais les effets sur les rendements ne sont pas significatifs. En tout cas, les plants de niébé réagissent mieux que les plants de petit mil, avec un rendement légèrement plus important dans des champs labourés que dans des champs non travaillés.

(ii) Des résidus de récoltes ont été appliqués à raison de 500 et de 2000 kg ha^{-1}. Un paillage de résidus de récoltes à raison de 2000 kg ha^{-1} réduit la vitesse du vent et par conséquent, le flux du sol. Cependant, un paillage de 500 kg ha^{-1} n'a pas d'effet. Le seuil quantitatif de résidus de récoltes nécessaires pour réduire l'érosion s'avère se situer entre 800 et 2000 kg ha^{-1}, sans qu'on le connaisse toutefois de façon précise. Seule la quantité maximale de 2000 kg ha^{-1} de résidus de récoltes a des effets significatifs sur les rendements des cultures. Ceci non seulement en raison de la réduction nette des flux du sol et d'une déposition de sol supplémentaire dans les résidus de récoltes, mais aussi en raison d'un apport d'éléments nutritifs provenant du paillage. Cependant, la quantité de résidus récoltés lors d'une année normale est inférieure à 2000 kg ha^{-1} rendant impossible une application de résidus de récoltes en une telle quantité. Compte tenu de la valeur des résidus de récoltes de petit mil en tant que fourrage et ses multiples usages domestiques, son application par les agriculteurs pour réduire l'érosion éolienne peut être limitée à des parties du champ sévèrement atteintes par celle-ci. Concentrer les résidus de récoltes disponibles sur les zones à faible productivité constitue une stratégie à court terme.

(iii) Organiser une végétation de savane naturelle en bandes perpendiculaires à la direction principale du vent et cultiver entre celles-ci, permet de créer un système de brise-vent qui protège la récolte pour un coût relativement faible. Durant les trois années de recherche, on constate qu'un effet de maîtrise de l'érosion éolienne n'est obtenu qu'entre les bandes espacées de 6 à 20 m de distance. Etant donné la hauteur des brise-vent observée lors de la première année (0,6 m), la protection des cultures avec des espacements plus grands n'a pas été possible. La croissance du petit mil et du niébé est affectée de manière positive mais seulement durant leur période précoce de développement. A des stades ultérieurs de croissance, cet avantage disparaît et peut même devenir un désavantage en raison probablement de la compétition pour l'eau et les éléments nutritifs entre les cultures et les brise-vent étroitement espacés.

(iv) Les arbres et les herbes *Adropogon gayanus* plantés en rangées de brise-vent ont des effets mesurables sur l'érosion éolienne. Les brise-vent réduisent le flux du sol et produisent des effets divers en fonction de leur espèce. Après cinq années on peut mesurer un changement des propriétés du sol seulement à proximité étroite des brise-vent. Les résultats de ces expériences démontrent que jusqu'à une distance des brise-vent égalant dix fois leur hauteur, la croissance du petit mil réagit favorablement aux brise-vent. De plus, certaines espèces comme le *Faidherbia albida* augmentent les rendements en grain et en matière sèche de petit mil. Cependant, même ces arbres de brise-vent peuvent ne pas justifier leur coût, à moins que les valeurs des sous-produits tels que le bois et le fourrage qu'ils génèrent ne soient suffisantes.

Analyses économiques et point de vue des agriculteurs sur les techniques alternatives

L'évaluation de méthodes alternatives visant à réduire l'érosion éolienne et à améliorer les rendements des cultures démontre le potentiel prometteur des résidus de récoltes et des brise-vent en termes techniques. Afin de définir plus avant ce potentiel, une évaluation en termes de performance financière et économique a été entreprise. On a aussi mené une enquête sur une exploitation prenant en compte les opinions des agriculteurs sur ces options. L'évaluation économique a été basée sur les bénéfices directs (les produits des arbres) et indirects (la protection offerte par les brise-vent). Les résultats montrent que la valeur économique des différentes parties des arbres – bois et autres – ainsi que d'autres plantes utilisées comme brise-vent, modifient considérablement les retombées des brise-vent. C'est pourquoi il est important d'inclure ces éléments dans une analyse économique. Dans l'analyse présente, le manque d'informations n'a pas permis d'inclure les bénéfices fournis par les gousses de *Faidherbia albida* en tant qu'aliment, ceux de la gomme arabique pour l'*Acacia senegal*, l'écorce de l'*Acacia nilotica* pour le tannage du cuir, ceux du bois pour l'*Azadirachta indica*, ceux du fourrage de l'*Andropogon gayanus*, et les propriétés médicinales de toutes les espèces. De plus, lorsqu'on prévoit de planter des brise-vent on doit proposer une diversité d'espèces. A l'échelle d'une unité domestique de production les applications de résidus de récoltes sont très prometteuses même si la force de travail est un facteur limitant. L'opinion des agriculteurs concernant les arbres et les applications de résidus de récoltes ont confirmé les résultats des analyses financières et économiques effectuées dans le cadre de cette étude. Le schéma qui se dégage de cette évaluation est le suivant: (i) de tous les arbres, le *Faidherbia albida* est le plus intéressant dans l'usage de brise-vent, même non plantés en rangées, (ii) l'application des résidus de récoltes doit s'effectuer essentiellement sur les parties des champs à faible productivité. Cependant, comme il est probable qu'à l'avenir la terre sera de plus en plus rare, le paillage sera compétitif avec les usages traditionnels des résidus de récoltes pour la maîtrise de l'érosion éolienne.

Perspectives et recommendations

Les besoins de la recherche à venir

La plupart des recherches menées jusqu'ici en station en Afrique de l'Ouest sahélienne ont généré d'excellentes techniques visant à réduire la dégradation des sols et à augmenter leur fertilité. Cependant ces techniques n'ont pas été acceptées sans réserves dans les systèmes de production locaux. Les ménages du Sahel ont de nombreux objectifs et entreprennent des activités variées afin de les atteindre. L'agriculture, l'élevage, le commerce, l'art, et les travaux d'appoint à l'étranger représentent quelques-unes des activités réparties sur l'année et qui s'inscrivent dans des perspectives à long, moyen ou court terme. Les multiples intérêts économiques des ménages ruraux laissent peu de force de travail disponible pour l'implantation d'innovations, en particulier pour lutter contre l'érosion éolienne. C'est pourquoi toute technologie recommandée doit être suffisamment flexible pour s'adapter à la diversité et la variabilité des systèmes de production de la région. Afin de permettre cela, des enquêtes socio-économiques et socio-culturelles doivent être intégrées aux expériences technologiques, nécessitant une approche interdisciplinaire. De plus, on doit généraliser la participation systématique des agriculteurs à l'évaluation et à l'adaptation des innovations.

Des outils analytiques pour une étude sur le long terme restent à développer et requièrent une recherche approfondie. Ces outils doivent incorporer la dimension temporelle, prendre en compte les dynamiques de facteurs tels que l'érosion éolienne dans les systèmes de production, et extrapoler les résultats obtenus au niveau de l'exploitation à une plus grande échelle, régionale ou nationale. Ils doivent aussi permettre de quantifier et évaluer par ailleurs les dommages et améliorations sur l'environnement. Aussi, afin de prendre en compte les facteurs de risque et d'aléa qui influencent les agriculteurs dans leurs décisions d'adopter des techniques de lutte contre l'érosion éolienne, on a besoin d'informations sur le long terme, quantifiant la relation entre l'érosion éolienne et les brise-vent (ou toute autre technique anti-érosive en fait) ainsi que les retombées financières différées.

Les chercheurs en sciences sociales en appellent au développement d'outils analytiques dans les domaines de la science des sols et de l'économie pour évaluer les différents degrés de l'érosion de la terre parce que les décisions d'attribuer des ressources limitées pour le contrôle de l'érosion ne peuvent être basées sur les seules analyses bénéfice/coût.

Les critères socialement pertinents, comme la durabilité des systèmes de production, l'équilibre régional et autre effets d'équité influencés par l'érosion du sol n'ont pas été pris en compte du point de vue des paysans. Des analyses sociales doivent par conséquent être inclues lorsqu'il s'agit de prendre des décisions à l'échelle nationale ou même régionale. Le choix d'un taux d'escompte social, par rapport à un taux d'escompte commercial pour le capital, pose des difficultés particulières dans le contexte de l'évaluation de phénomènes à aussi long terme que l'érosion des sols.

Afin de se concentrer sur l'implantation possible de différentes techniques de contrôle de l'érosion des sols, les recherches futures doivent aussi être dirigées vers des investigations régionales et la caractérisation des zones affectées par l'érosion éolienne. On dispose à cet effet d'outils importants tels que la photographie aérienne à faible coût (Buerkert *et al.* 1995), ou des images satellitaires déterminant l'utilisation de la terre et la végétation. On dispose également de systèmes d'information géographique qui améliorent l'interprétation de ces informations (Quiroz *et al.* 1995). Un aboutissement intéressant de ces recherches pourrait consister en une carte régionale de probabilité de l'érosion éolienne indiquant les techniques idéales et possibles de la maîtrise de cette érosion mises en relation avec les différentes cultures importantes pour la région.

Comme le petit mil est la nourriture de base de la région, beaucoup de recherches ont été axées sur cette culture. Cependant, l'augmentation de la production de petit mil en elle-même peut ne pas suffire à justifier des innovations coûteuses, en raison d'un prix de marché du petit mil généralement faible. D'autres recherches sont nécessaires sur d'autres cultures d'une valeur économique supérieure telles que le niébé et l'arachide.

Recommendations sur le plan politique et institutionnel

Pour que les mesures de lutte contre l'érosion éolienne basées sur la plantation d'arbres soient acceptées par les populations rurales elles doivent de préférence être simples; basées sur le savoir local; bon marché; sans besoins de maintenance, d'outils spéciaux ou de méthodes d'arrosage supplémentaires; rentables à court terme et gérables par les membres d'un ménage afin de minimiser l'organisation communautaire. Même dans ces conditions, de telles mesures exigent encore un soutien institutionnel et politique pour leur mise en place et le coût de ce support ne doit pas être sous-estimé.

Un critère important de succès dans l'implantation de ces techniques de lutte contre l'érosion des sols est l'acceptation totale de la responsabilité de la gestion des ressources naturelles par les populations rurales qui en sont les bénéficiaires. La documentation sur le droit foncier insiste sur le fait qu'en raison du statut précaire des droits fonciers, les propriétaires sont réticents à investir dans des techniques de conservation du sol, particulièrement dans les investissements dont les retombées n'interviennent qu'à long terme comme ceux des brise-vent (Kerkhof 1990; Neef *et al.* 1996). Le droit foncier, la propriété des terres et des arbres et autres droits doivent être clarifiés avant que les activités de lutte contre l'érosion ne soient entreprises à une grande échelle. Dans ce contexte, la propriété individuelle peut ne pas nécessairement être la seule réponse; la conscience de la nécessité de protéger l'environnement et de conserver les sols autant que la motivation individuelle pour entreprendre de telles activités, peuvent aussi être développées sur les propriétés communes. Cette conscience et cette motivation sont essentielles pour garantir l'installation de toute technique anti-érosive. Motiver les agriculteurs à utiliser ces innovations qui dégagent tout d'abord des profits à long terme est vital, car cela leur permettrait de subvenir aux besoins de leurs familles et de préserver leur ressources de base. La motivation des agriculteurs est une tâche difficile de vulgarisation, mais on doit aussi s'assurer d'un soutien politique et institutionnel. Ceci nécessite la motivation de la communauté toute entière et les décisions prises par la communauté doivent l'être par des représentants de tous les groupes intéressés de la société.

Les fermiers doivent être intégrés non seulement durant les phases de mise en place et d'entretien de la plantation des arbres, mais aussi lors des phases d'identification et de planification. Le programme mis en place ne doit pas s'appliquer aux seuls champs individuels mais aussi à la gestion des terres du village. Le personnel de vulgarisation doit être encouragé à jouer le rôle d'un intermédiaire qui guide la communauté dans ses choix en évitant de lui donner une opinion sur des sujets précis ou d'influencer la prise de décision.

La durée nécessaire à convaincre et à mobiliser les communautés ne doit pas être sous-estimée car elle peut se compter parfois en année.

Etant donnée l'importance des programmes de vulgarisation, l'organisation de la vulgarisation à l'échelle individuelle doit être prise en compte lorsqu'on définit les activités de vulgarisation des brise-vent. La plantation d'arbres et de brise-vent relève normalement de la responsabilité du département d'agro-foresterie. Cependant, dans de nombreux pays du Sahel, ces ministères sont structurés comme des organisations para-militaires, avec d'une part, une fonction de vulgarisation, d'autre part des fonctions de police et de contrôle (Lamers 1995). A cause de la nature duale et contradictoire de leurs attributions, le succès de ces agences est infime. Il faudrait réévaluer les objectifs politiques et institutionnels de la plupart des gouvernements locaux pour que toute installation de techniques de conservation des sols aboutisse.

Les sols sujets à l'érosion perdent des éléments nutritifs importants. Si une parcelle protégée est cultivée, les augmentations des rendements sont assurées grâce à la protection du petit mil par les brise-vent, mais cela augmente aussi l'exportation d'éléments nutritifs des sols vers les produits récoltés. Si ces éléments nutritifs ne sont pas remplacés, l'appauvrissement du sol sera plus rapide que sur une parcelle sans brise-vent. D'un point de vue agronomique, l'usage d'engrais organiques et minéraux est impératif. Les agriculteurs sahéliens les appliquent souvent en quantités insuffisantes en raison de la faible disponibilité des engrais organiques, de leur coût élevé et de l'inaccessibilité des engrais minéraux. Il leur arrive aussi de ne pas utiliser du tout d'engrais minéraux en raison de l'improbabilité des pluies ou d'un manque de capitaux. L'accessibilité et la distribution d'intrants bon marchés doivent donc être assurées et exigent une intervention immédiate des responsables politiques. Sans ces engrais minéraux, l'érosion au Sahel peut difficilement être réduite.

Le contrôle de l'érosion des sols est une entreprise qui dépasse de beaucoup le cadre d'une exploitation et même d'une nation sahélienne. Si l'on n'arrête pas l'érosion maintenant, la population sahélienne qui a tant bien que mal survécu jusqu'à aujourd'hui, devra à l'avenir importer des denrées alimentaires à des prix excessivement élevés. Cette importation dépasserait de beaucoup, selon toute probabilité les coûts exigés aujourd'hui par la mise en place de technologies de conservation des sols. Retarder leur mise en place est un sujet de débat parce que les coûts d'opportunité des revenus des agriculteurs sont élevés. A l'échelle mondiale, les techniques de lutte contre l'érosion éolienne n'ont jamais été implantées à grande échelle sans des motivations financières et économiques provenant de sources extérieures à l'exploitation (Lamers *et al.* 1996). C'est pourquoi il n'est pas réaliste d'attendre des agriculteurs sahéliens qu'ils adoptent ces mesures d'eux-mêmes sans aide financière. De même, des incitations économiques et une politique agricole des prix favorable sont requis pour renforcer l'impact de la recherche et de la vulgarisation. L'instauration de marchés plus nombreux pour le bois de feu et l'alimentation peut contribuer à faciliter l'adoption de brise-vent d'arbres plantés en rangs, ou d'arbres isolés dans une région à la végétation clairsemée.

Au vu des activités innombrables qui se tiennent dans et autour de ce réservoir de poussière qu'est le Sahara, la possibilité d'un effet des interventions humaines individuelles sur la stabilisation de la fertilité des sols peut paraître désespérément faible. Cependant, il est crucial d'un point de vue national et même plus global de développer des systèmes de production durables au Sahel. Si l'on considère par conséquent que la tâche de maintenir la fertilité des sols relève d'une responsabilité à l'échelle internationale, on trouvera les moyens pour la mise en place de ces mesures anti-érosives. Les techniques décrites dans ce livre peuvent ne pas avoir la capacité de résoudre le problème de l'érosion éolienne dans son ensemble. Cependant, elles existent et peuvent être efficaces à certaines conditions. Ainsi le défi consiste à aller de l'avant quant à leur utilisation qui se doit d'être pertinente et basée sur l'effort individuel mais aussi sur un soutien national et international.

Bibliographie

Buerkert, A., F. Mahler, and H. Marschner. 1996. Soil productivity management and plant growth in the Sahel: Potential of an aerial monitoring technique. *Plant and Soil* 180(1): 29–38.

Kerkhof, P. 1990. Agroforestry in Africa: A survey of project experience. pp. 113–123. *In* G. Foley and G. Barnard (eds.) Panos Technical Series. Panos Publications London Ltd.

Lamers, J.P.A. 1995. An assessment of wind erosion control techniques in Niger: Financial and economic analyses of windbreaks and millet crop residues. Diss. Universität Hohenheim, Verlag Ulrich E. Grauer, Stuttgart, Germany.

Lamers, J.P.A., T. Becker, and M. von Oppen. 1996. Arrangements to finance tree windbreaks for wind erosion control. pp. 227–241. *In* B. Buerkert, B.E. Allison, and M. von Oppen (eds.) Wind erosion in West Africa: The problem and its control. Proceedings International Symposium, Hohenheim, Germany, 5–7 Dec. 1994, Margraf Verlag, Weikersheim, Germany.

Neef A., J. Haigis, and F. Heidhues. 1996. L'impact de pluralisme institutionnel et juridique sur l'indroduction de mesures anti-érosives – le cas du Niger. pp. 283–296. *In* B. Buerkert, B.E. Allison, and M. von Oppen (eds.) Wind erosion in West Africa: The problem and its control. Proceedings International Symposium, Hohenheim, Germany, 5–7 Dec. 1994, Margraf Verlag, Weikersheim, Germany.

Quiroz, R.A., R.B. Estrada, C.L. Velarde, and H.G. Zandstra. 1995. The use of simulation models in developing sustainable management of natural resources for the Andean zone. *International Potato Center (CIP) Circular* 21(1): 10–18.

Developments in Plant and Soil Sciences

1. J. Monteith and C. Webb (eds.): *Soil Water and Nitrogen in Mediterranean-type Environments*. 1981 ISBN 90-247-2406-6
2. J.C. Brogan (ed.): *Nitrogen Losses and Surface Run-off from Landspreading of Manures*. 1981 ISBN 90-247-2471-6
3. J.D. Bewley (ed.): *Nitrogen and Carbon Metabolism*. 1981 ISBN 90-247-2472-4
4. R. Brouwer, I. Gašparíková, J. Kolek and B.C. Loughman (eds.): *Structure and Function of Plant Roots*. 1981 ISBN 90-247-2510-0
5. Y.R. Dommergues and H.G. Diem (eds.): *Microbiology of Tropical Soils and Plant Productivity*. 1982 ISBN 90-247-2624-7
6. G.P. Robertson, R. Herrara and T. Rosswall (eds.): *Nitrogen Cycling in Ecosystems of Latin America and the Caribbean*. 1982 ISBN 90-247-2719-7
7. D. Atkinson, K.K.S. Bhat, M.P. Coutts, P.A. Mason and D.J. Read (eds.): *Tree Root Systems and Their Mycorrhizas*. 1983 ISBN 90-247-2821-5
8. M.R. Sarić and B.C. Loughman (eds.): *Genetic Aspects of Plant Nutrition*. 1983 ISBN 90-247-2822-3
9. J.R. Freney and J.R. Simpson (eds.): *Gaseous Loss of Nitrogen from Plant-Soil Systems*. 1983 ISBN 90-247-2820-7
10. United Nations Economic Commission for Europe (ed.): *Efficient Use of Fertilizers in Agriculture*. 1983 ISBN 90-247-2866-5
11. J. Tinsley and J.F. Darbyshire (eds.): *Biological Processes and Soil Fertility*. 1984 ISBN 90-247-2902-5
12. A.D.L. Akkermans, D. Baker, K. Huss-Danell and J.D. Tjepkema (eds.): Frankia *Symbioses*. 1984 ISBN 90-247-2967-X
13. W.S. Silver and E.C. Schröder (eds.): *Practical Application of* Azolla *for Rice Production*. 1984 ISBN 90-247-3068-6
14. P.G.L. Vlek (ed.): *Micronutrients in Tropical Food Crop Production*. 1985 ISBN 90-247-3085-6
15. T.P. Hignett (ed.): *Fertilizer Manual*. 1985 ISBN 90-247-3122-4
16. D. Vaughan and R.E. Malcolm (eds.): *Soil Organic Matter and Biological Activity*. 1985 ISBN 90-247-3154-2
17. D. Pasternak and A. San Pietro (eds.): *Biosalinity in Action*. Bioproduction with Saline Water. 1985 ISBN 90-247-3159-3
18. M. Lalonde, C. Camiré and J.O. Dawson (eds.): Frankia *and Actinorhizal Plants*. 1985 ISBN 90-247-3214-X
19. H. Lambers, J.J. Neeteson and I. Stulen (eds.): *Fundamental, Ecological and Agricultural Aspects of Nitrogen Metabolism in Higher Plants*. 1986 ISBN 90-247-3258-1
20. M.B. Jackson (ed.): *New Root Formation in Plants and Cuttings*. 1986 ISBN 90-247-3260-3
21. F.A. Skinner and P. Uomala (eds.): *Nitrogen Fixation with Non-Legumes* (Proceedings of the 3rd Symposium, Helsinki, 1984). 1986 ISBN 90-247-3283-2
22. A. Alexander (ed.): *Foliar Fertilization*. 1986 ISBN 90-247-3288-3
23. H.G. v.d. Meer, J.C. Ryden and G.C. Ennik (eds.): *Nitrogen Fluxes in Intensive Grassland Systems*. 1986 ISBN 90-247-3309-X
24. A.U. Mokwunye and P.L.G. Vlek (eds.): *Management of Nitrogen and Phosphorus Fertilizers in Sub-Saharan Africa*. 1986 ISBN 90-247-3312-X
25. Y. Chen and Y. Avnimelech (eds.): *The Role of Organic Matter in Modern Agriculture*. 1986 ISBN 90-247-3360-X

Developments in Plant and Soil Sciences

26. S.K. De Datta and W.H. Patrick Jr. (eds.): *Nitrogen Economy of Flooded Rice Soils.* 1986 ISBN 90-247-3361-8
27. W.H. Gabelman and B.C. Loughman (eds.): *Genetic Aspects of Plant Mineral Nutrition.* 1987 ISBN 90-247-3494-0
28. A. van Diest (ed.): *Plant and Soil: Interfaces and Interactions.* 1987 ISBN 90-247-3535-1
29. United Nations Economic Commission for Europe and FAO (eds.): *The Utilization of Secondary and Trace Elements in Agriculture.* 1987 ISBN 90-247-3546-7
30. H.G. v.d. Meer, R.J. Unwin, T.A. van Dijk and G.C. Ennik (eds.): *Animal Manure on Grassland and Fodder Crops.* Fertilizer or Waste? 1987 ISBN 90-247-3568-8
31. N.J. Barrow: *Reactions with Variable-Charge Soils.* 1987 ISBN 90-247-3589-0
32. D.P. Beck and L.A. Materon (eds.): *Nitrogen Fixation by Legumes in Mediterranean Agriculture.* 1988 ISBN 90-247-3624-2
33. R.D. Graham, R.J. Hannam and N.C. Uren (eds.): *Manganese in Soils and Plants.* 1988 ISBN 90-247-3758-3
34. J.G. Torrey and J.L. Winship (eds.): *Applications of Continuous and Steady-State Methods to Root Biology.* 1989 ISBN 0-7923-0024-6
35. F.A. Skinner, R.M. Boddey and I. Fendrik (eds.): *Nitrogen Fixation with Non-Legumes* (Proceedings of the 4th Symposium, Rio de Janeiro, 1987). 1989 ISBN 0-7923-0059-9
36. B.C. Loughman, O. Gašparíková and J. Kolek (eds.): *Structural and Functional Aspects of Transport in Roots.* 1989 ISBN 0-7923-0060-2; Pb 0-7923-0061-0
37. P. Plancquaert and R. Haggar (eds.): *Legumes in Farming Systems.* 1990 ISBN 0-7923-0134-X
38. A.E. Osman, M.M. Ibrahim and M.A. Jones (eds.): *The Role of Legumes in the Farming Systems of the Mediterranean Areas.* 1990 ISBN 0-7923-0419-5
39. M. Clarholm and L. Bergström (eds.): *Ecology of Arable Land – Perspectives and Challenges.* 1989 ISBN 0-7923-0424-1
40. J. Vos, C.D. van Loon and G.J. Bollen (eds.): *Effects of Crop Rotation on Potato Production in the Temperate Zones.* 1989 ISBN 0-7923-0495-0
41. M.L. van Beusichem (ed.): *Plant Nutrition – Physiology and Applications.* 1990 ISBN 0-7923-0740-2
42. N. El Bassam, M. Dambroth and B.C. Loughman (eds.): *Genetic Aspects of Plant Mineral Nutrition.* 1990 ISBN 0-7923-0785-2
43. Y. Chen and Y. Hadar (eds.): *Iron Nutrition and Interactions in Plants.* 1991 ISBN 0-7923-1095-0
44. J.J.R. Groot, P. de Willigen and E.L.J. Verberne (eds.): *Nitrogen Turnover in the Soil-Crop System.* 1991 ISBN 0-7923-1107-8
45. R.J. Wright, V.C. Baligar and R.P. Murrmann (eds.): *Plant-Soil Interactions at Low pH.* 1991 ISBN 0-7923-1105-1
46. J. Kolek and V. Kozinka (eds.): *Physiology of the Plant Root System.* 1992 ISBN 0-7923-1205-8
47. A.U. Mokwunye (ed.): *Alleviating Soil Fertility Constraints to Increased Crop Production in West Africa.* 1991 ISBN 0-7923-1221-X; Pb 0-7923-1222-8
48. M. Polsinelli, R. Materassi and M. Vincenzini (eds.): *Nitrogen Fixation* (Proceedings of the 5th Symposium, Florence, 1990). 1991 ISBN 0-7923-1410-7
49. J.K. Ladha, T. George and B.B. Bohlool (eds.): *Biological Nitrogen Fixation for Sustainable Agriculture.* 1992 ISBN 0-7923-1774-2

Developments in Plant and Soil Sciences

50. P.J. Randall, E. Delhaze, R.A. Richards and R. Munns (eds.): *Genetic Aspects of Plant Mineral Nutrition.* 1993 ISBN 0-7923-2118-9
51. K.S. Kumarasinghe and D.L. Eskew (eds.): *Isotopic Studies of Azolla and Nitrogen Fertilization of Rice.* 1993 ISBN 0-7923-2274-6
52. F.A. Bliss and G. Hardarson (eds.): *Enhancement of Biological Nitrogen Fixation of Common Baen in Latin America.* 1993 ISBN 0-7923-2451-X
53. M.A.C. Fragoso and M.L. van Beusichem (eds.): *Optimization of Plant Nutrition.* 1993 ISBN 0-7923-2519-2
54. N.J. Barrow (ed.) *Plant Nutrition – From Genetic Engineering to Field Practice.* 1993 ISBN 0-7923-2540-0
55. A.D. Robson (ed.): *Zinc in Soils and Plants.* 1993 ISBN 0-7923-2631-8
56. A.D. Robson, L.K. Abbott and N. Malajczuk (eds.): *Management of Mycorrhizas in Agriculture, Horticulture and Forestry.* 1994 ISBN 0-7923-2700-4
57. P.H. Graham, M.J. Sadowsky and C.P. Vance (eds.): *Symbiotic Nitrogen Fixation.* 1994 ISBN 0-7923-2781-0
58. F. Baluška, M. Čiamporová, O. Gašparíkova and P.W. Barlow (eds.): *Structure and Function of Roots.* 1995 ISBN 0-7923-2832-9
59. J. Abadía (ed.): *Iron Nutrition in Soils and Plants.* 1995 ISBN 0-7923-2900-7
60. P.S. Curtis, E.G. O'Neill, J.A. Teeri, D.R. Zak and K.S. Pregitzer (eds.): *Belowground Responses to Rising Atmospheric CO_2.* Implications for Plants, Soil Biota, and Ecosystem Processes. 1994 ISBN 0-7923-2901-5
61. P.C. Struik, W.J. Vredenberg, J.A. Renkema and J.E. Parlevliet (eds.): *Plant Production on the Threshold of a New Century.* 1994 ISBN 0-7923-2903-1
62. L.O. Nilsson, R.F. Hüttl and U.T. Johansson (eds.): *Nutrient Uptake and Cycling in Forest Ecosystems* (CEC/IUFRO Symposium, Halmstad, Sweden, 1993). 1995 ISBN 0-7923-3030-7
63. H.P. Collins, G.P. Robertson and M.J. Klug (eds.): *The Significance and Regulation of Soil Biodiversity* (International Symposium, East Lansing, 1993). 1995 ISBN 0-7923-3138-9
64. R.A. Date, N.J. Grundon, G.E. Rayment and M.E. Probert (eds.): *Plant-Soil Interactions at Low pH: Principles and Management* (3rd International Symposium, Brisbane, 1993). 1995 ISBN 0-7923-3198-2
65. J.K. Ladha and M.B. Peoples (eds.): *Management of Biological Nitrogen Fixation for the Development of More Productive and Sustainable Agricultural Systems.* (15th Congress of Soil Science, Acapulco, Mexico, 1994) 1995 ISBN 0-7923-3413-2; Pb 0-7923-3414-0
66. C. Rodriguez-Barrueco (ed.): *Fertilizers and Environment.* 1996 ISBN 0-7923-3729-8
67. B. Buerkert, B.E. Allison and M. von Oppen (eds.): *Wind Erosion in Niger.* Implications and Control Measures in a Millet-based Farming System. 1996 ISBN 0-7923-3885-5
68. O. van Cleemput, G. Hofman and A. Vermoesen (eds.): *Progress in Nitrogen Cycling.* 1996 ISBN 0-7923-3962-2
69. N. Ahmad (ed.): *Nitrogen Economy in Tropical Soils.* 1996 ISBN 0-7923-4094-9

Kluwer Academic Publishers – Dordrecht / Boston / London

FSC
www.fsc.org
MIX
Papier aus verantwortungsvollen Quellen
Paper from responsible sources
FSC® C105338

If you have any concerns about our products,
you can contact us on
ProductSafety@springernature.com

In case Publisher is established outside the EU,
the EU authorized representative is:
Springer Nature Customer Service Center GmbH
Europaplatz 3, 69115 Heidelberg, Germany

Printed by Libri Plureos GmbH
in Hamburg, Germany